AQA
GCSE science

D0189032

Authors

Graham Bone

Simon Broadley

Philippa Gardom Hulme

Sue Hocking

Mark Matthews

Jim Newall

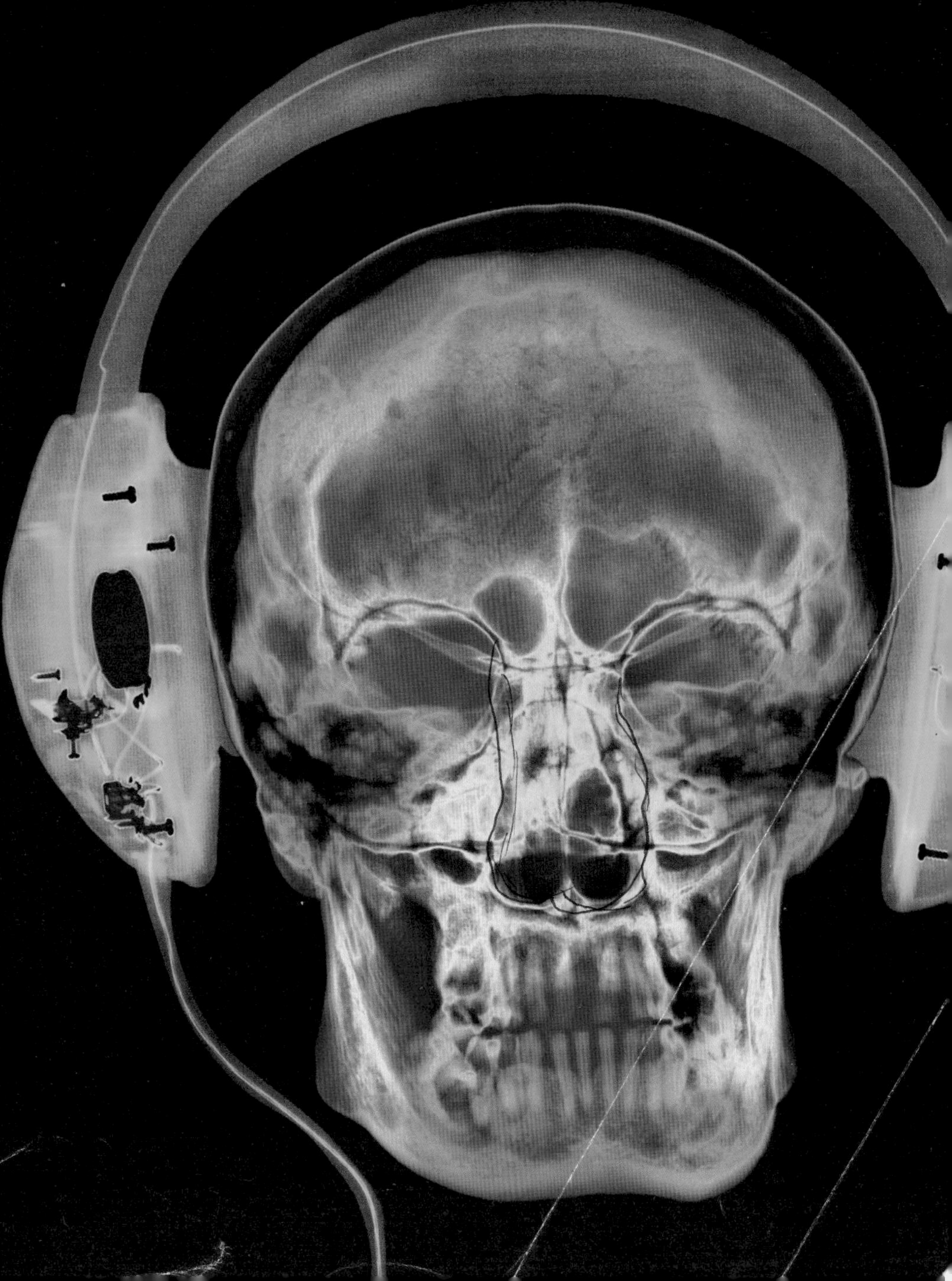

Contents

How to use this book

Welcome to your AQA GCSE Science A course. This book has been specially written by experienced teachers and examiners to match the 2011 specification.

On these two pages you can see the types of pages you will find in this book, and the features on them. Everything in the book is designed to provide you with the support you need to help you prepare for your examinations and achieve your best.

Unit openers

Specification matching grid: This shows you how the pages in the unit match to the exam specification for GCSE Science, so you can track your progress through the unit as you learn.

Why study this unit: Here you can read about the reasons why the science you're about to learn is relevant to your everyday life.

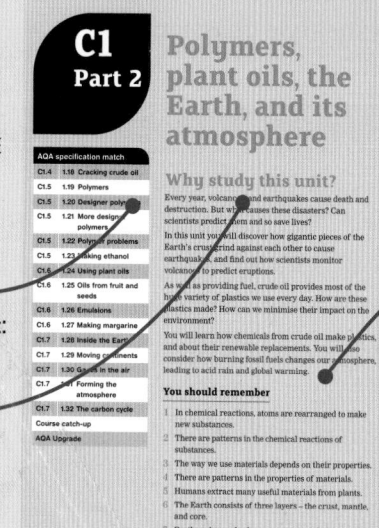

You should remember: This list is a summary of the things you've already learnt that will come up again in this unit. Check through them in advance and see if there is anything that you need to recap on before you get started.

Opener image: Every unit starts with a picture and information on a new or interesting piece of science that relates to what you're about to learn.

Main pages

Learning objectives: You can use these objectives to understand what you need to learn to prepare for your exams. Higher Tier only objectives appear in pink text.

Key words: These are the terms you need to understand for your exams. You can look for these words in the text in bold or check the glossary to see what they mean.

Questions: Use the questions on each spread to test yourself on what you've just read.

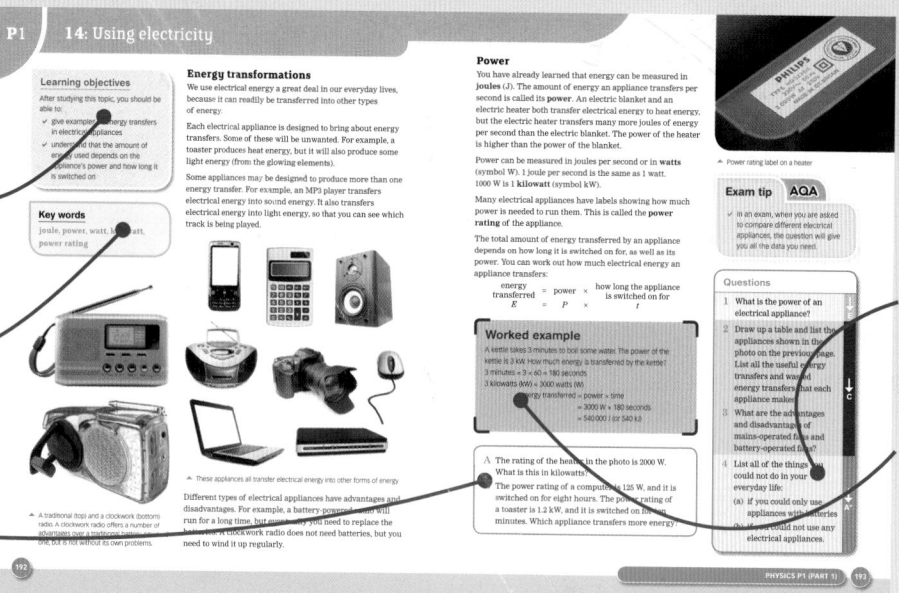

Higher Tier content: Anything marked in pink is for students taking the Higher Tier paper only. As you go through you can look at this material and attempt it to help you understand what is expected for the Higher Tier.

Worked examples: These help you understand how to use an equation or to work through a calculation. You can check back whenever you use the calculation in your work.

Summary and exam-style questions

Every summary question at the end of a spread includes an indication of how hard it is. These indicators show which grade you are working towards. You can track your own progress by seeing which of the questions you can answer easily, and which you have difficulty with.

When you reach the end of a unit you can use the exam-style questions to test how well you know what you've just learnt. Each question has a grade band next to it.

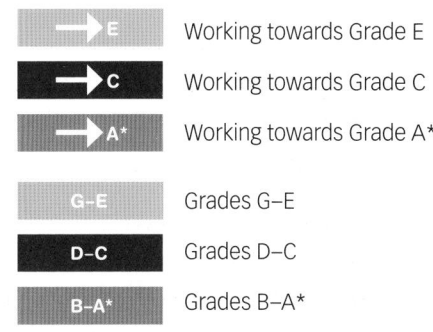

E — Working towards Grade E
C — Working towards Grade C
A* — Working towards Grade A*

G–E — Grades G–E
D–C — Grades D–C
B–A* — Grades B–A*

Course catch-ups

Revision checklist: This is a summary of the main ideas in the unit. You can use it as a starting point for revision, to check that you know about the big ideas covered.

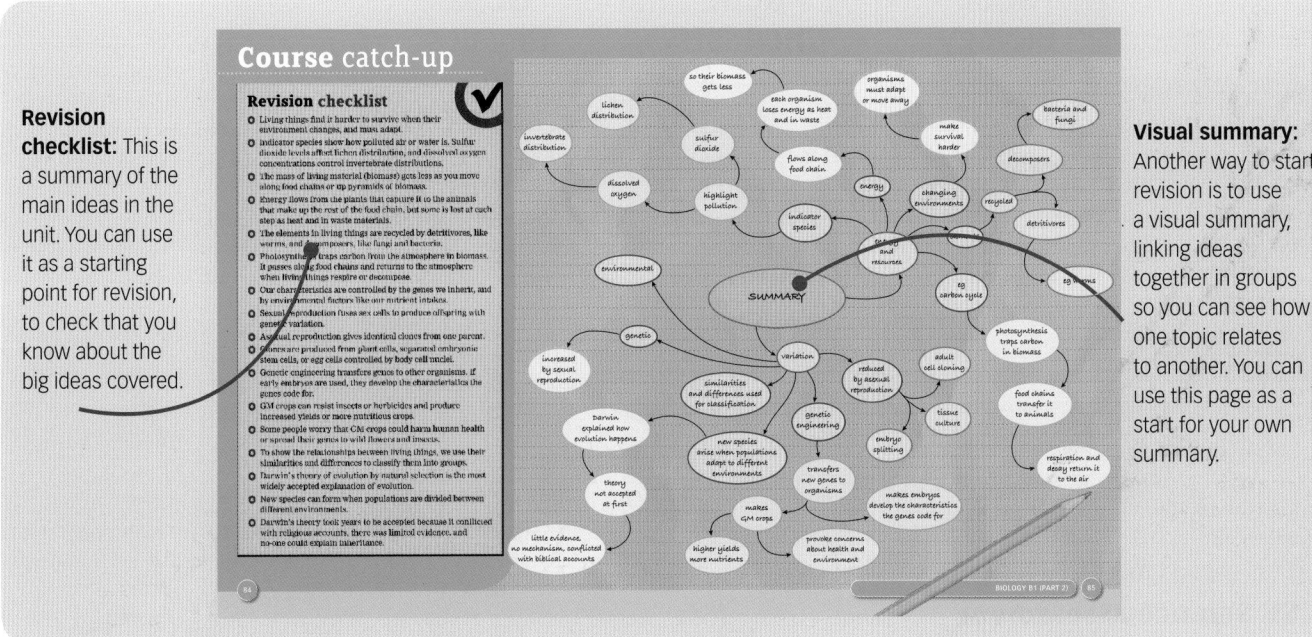

Visual summary: Another way to start revision is to use a visual summary, linking ideas together in groups so you can see how one topic relates to another. You can use this page as a start for your own summary.

Upgrade: Upgrade takes you through an exam question in a step-by-step way, showing you why different answers get different grades. Using the tips on the page you can make sure you achieve your best by understanding what each question needs.

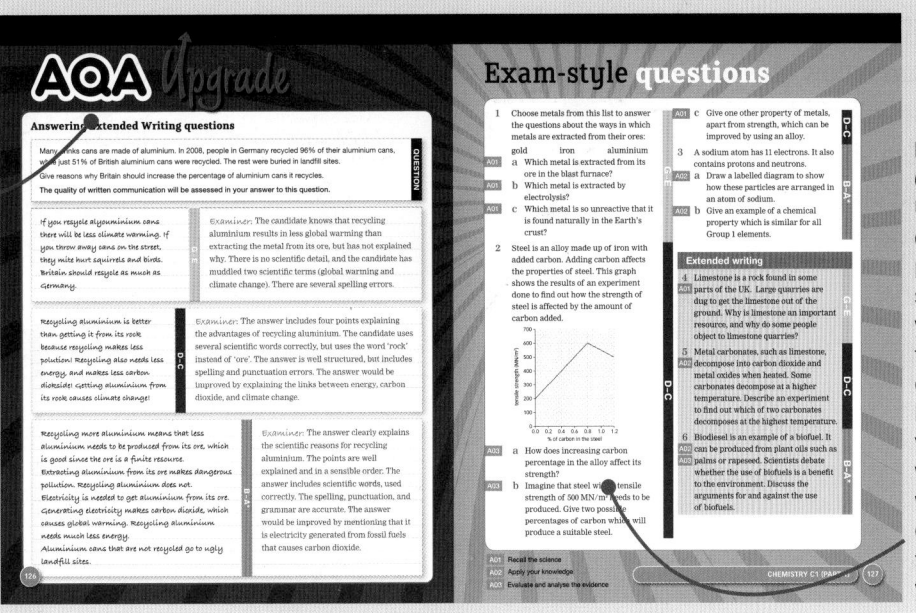

Exam-style questions: Using these questions you can practice your exam skills, and make sure you're ready for the real thing. Each question has a grade band next to it, so you can understand what level you are working at and focus on where you need to improve to get your target grade.

AQA Upgrade

Matching your course

The units in this book have been written to match the specification, no matter what you plan to study after your GCSE Science A course.

In the diagram below you can see that the units and part units can be used to study either for **GCSE Science**, leading to **GCSE Additional Science**, or as part of **GCSE Biology**, **GCSE Chemistry** and **GCSE Physics** courses.

	GCSE Biology	GCSE Chemistry	GCSE Physics
GCSE Science	B1 (Part 1)	C1 (Part 1)	P1 (Part 1)
	B1 (Part 2)	C1 (Part 2)	P1 (Part 2)
GCSE Additional Science	B2 (Part 1)	C2 (Part 1)	P2 (Part 1)
	B2 (Part 2)	C2 (Part 2)	P2 (Part 2)
	B3 (Part 1)	C3 (Part 1)	P3 (Part 1)
	B3 (Part 2)	C3 (Part 2)	P3 (Part 2)

GCSE Science assessment

The units in this book are broken into two parts to match the different types of exam paper on offer. The diagram below shows you what is included in each exam paper. It also shows you how much of your final mark you will be working towards in each paper.

	Unit		%	Type	Time	Marks available
Route 1	Unit 1	B1 (Part 1) / B1 (Part 2)	25%	Written exam	1 hr	60
	Unit 2	C1 (Part 1) / C1 (Part 2)	25%	Written exam	1 hr	60
	Unit 3	P1 (Part 1) / P1 (Part 2)	25%	Written exam	1 hr	60
	Unit 4	Controlled Assessment	25%		1 hr 30 mins + practical	50
Route 2	Unit 5	B1 (Part 1) / C1 (Part 1) / P1 (Part 1)	35%	Written exam	1 hr 30 mins	90
	Unit 6	B1 (Part 2) / C1 (Part 2) / P1 (Part 2)	40%	Written exam	1 hr 30 mins	90
	Unit 4	Controlled Assessment	25%		1 hr 30 mins + practical	50

Understanding exam questions — ภาพนอก

When you read the questions in your exam papers you should make sure you know what kind of answer you are being asked for. The list below explains some of the common words you will see used in exam questions. Make sure you know what each word means. Always read the question thoroughly, even if you recognise the word used. — can U understand ไป

Calculate

Work out your answer by using a calculation. You can use your calculator to help you. You may need to use an equation; check whether one has been provided for you in the paper. The question will say if your working must be shown.

Describe

Write a detailed answer that covers what happens, when it happens, and where it happens. The question will let you know how much of the topic to cover. Talk about facts and characteristics. (Hint: don't confuse with 'Explain')

point.

Explain

You will be asked how or why something happens. Write a detailed answer that covers how and why a thing happens. Talk about mechanisms and reasons. (Hint: don't confuse with 'Describe')

Evaluate

You will be given some facts, data or other information. Write about the data or facts and provide your own conclusion or opinion on them.

Outline

ให้เอาวย่อ

Give only the key facts of the topic. You may need to set out the steps of a procedure or process – make sure you write down the steps in the correct order.

Show

Write down the details, steps or calculations needed to prove an answer that you have been given.

Suggest

Think about what you've learnt in your science lessons and apply it to a new situation or a context. You may not know the answer. Use what you have learnt to suggest sensible answers to the question.

Write down

Give a short answer, without a supporting argument.

Top tips

Always read exam questions carefully, even if you recognise the word used. Look at the information in the question and the number of answer lines to see how much detail the examiner is looking for.

You can use bullet points or a diagram if it helps your answer.

If a number needs units you should include them, unless the units are already given on the answer line.

Controlled Assessment in GCSE Science

As part of the assessment for your GCSE Science A course, you will undertake a Controlled Assessment task.

What is Controlled Assessment?

Controlled Assessment has taken the place of coursework for the new 2011 GCSE Science specifications. The main difference between coursework and Controlled Assessment is that you will be supervised by your teacher when you carry out your Controlled Assessment task.

What will my Controlled Assessment task look like?

Your Controlled Assessment task will be made up of four sections. These four sections make up an investigation, with each section looking at a different part of the scientific process.

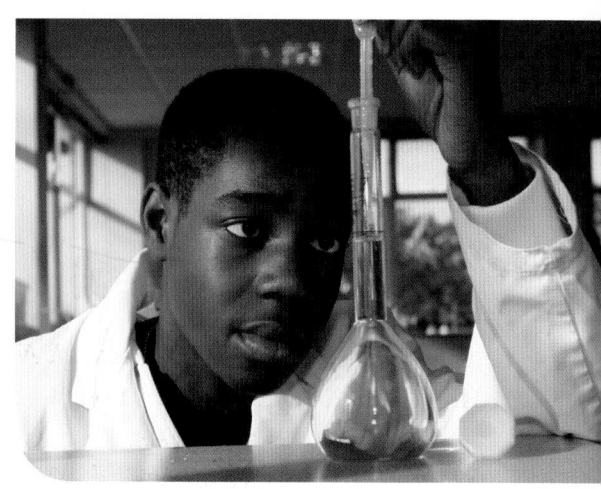

	What will I need to do?	How many marks are available?
Research	• Consider the hypothesis you have been given. • Research the method for carrying out an experiment to test the hypothesis. • Research the context of the investigation. • Carry out a risk assessment.	
Section 1	• Answer questions relating to your own research.	20 marks
Practical investigation	• Carry out your own experiment and record and analyse your results.	
Section 2	• Answer questions relating to the experiment you carried out. • Select appropriate data from data supplied by AQA and use it to analyse and compare with the hypothesis. • Suggest how your own hypothesis and research could be used within a new context.	30 marks
	Total	**50 marks**

How do I prepare for my Controlled Assessment?

Throughout your course you will learn how to carry out investigations in a scientific way, and how to analyse and compare data properly.

On the next three pages there are Controlled Assessment-style questions matched to the biology, chemistry, and physics content in B1, C1, and P1. You can use them to test yourself, and to find out which areas you want to practise more before you take the Controlled Assessment task itself.

B1 Controlled Assessment-style questions

Hypothesis: Light affects the distribution of woodlice. More woodlice are found in dark areas.

Download the Research Notes and Data Sheet for B1 from **www.oxfordsecondary.co.uk/ aqacasestudies**.

Research

*Record your findings in the **Research Notes table**.*

1. Research two different methods that could be used to test the hypothesis.
2. Find out how the results of the investigation might be useful in determining whether animals are well adapted to survive in their normal environment, and that organisms may sense and react to stimuli.

Section 1   Total 20 marks

Use your research findings to answer these questions.

1. (a) Name two sources that you used for your research.
   (b) Which of these sources did you find more useful, and why? [3]
2. (a) Identify one control variable.
   (b) Briefly describe a preliminary investigation to find a suitable value for this variable, and explain how the results of this work will help you decide on the best value for it. [3]
3. *In this question you will be assessed on using good English, organising information clearly, and using specialist terms where appropriate.* Describe how to carry out an investigation to test the hypothesis. Include the equipment needed and how to use it, the measurements to make, how to make it a fair test, and a risk assessment. [9]
4. Use your research to outline another possible method, and explain why you did not choose it. [3]
5. Draw a table to record data from the investigation. You may use ICT if you wish. [2]

Section 2   Total 30 marks

*Use the **Data Sheet** to answer these questions.*

1. (a) State the independent and dependent variables, and one control variable. [3]
   (b) Think about the time intervals for your data measurements. The smallest scale division on a measuring instrument is called its resolution. What was the resolution of the instrument you used, and was this resolution suitable for your experiment? [3]
   (c) Display the **Group A data** on a bar chart or line graph. *This data has been provided for you to use instead of data that you would gather yourself.* [4]
   (d) Does the **Group A data** support the hypothesis? Explain how. [3]
   (e) Describe the similarities and differences between the **Group A data** and the **Group B data**. Suggest one reason why the results of the two groups may be different. *The **Group B data** has been provided for you to use instead of data that would be gathered by others in your class.* [3]
   (f) Suggest how your method might have helped Group A to achieve results that show a clear pattern. [3]
2. (a) Sketch a graph of the results in **Case study 1**. [2]
   (b) Explain to what extent the data from **Case studies 1–3** support the hypothesis. [3]
   (c) Use **Case study 4** to describe the relationship between light and movement in woodlice. Explain how well the data supports your answer. [3]
3. The context of this investigation (the topic it relates to) is that animals are well adapted to survive in their normal environment, and that organisms may perceive and react to stimuli. Describe how your results may be useful in this context. [3]

C1 Controlled Assessment-style questions

Hypothesis: It is suggested that there is a link between the position of a metal in the reactivity series, and the ease of decomposition of its carbonate.

Download the Research Notes and Data Sheet for C1 from **www.oxfordsecondary.co.uk/aqacasestudies**.

Research

*Record your findings in the **Research Notes table**.*

1. Research two different methods that could be used to test the hypothesis.
2. Find out how the results of the investigation might be useful in determining how to extract metals from ores that contain metal carbonates.

Section 1   Total 20 marks

Use your research findings to answer these questions.

1. (a) Name two sources that you used for your research.
   (b) Which of these sources did you find more useful, and why?   [3]
2. (a) Identify one control variable.
   (b) Briefly describe a preliminary investigation to find a suitable value for this variable, and explain how the results of this work will help you decide on the best value for it.   [3]
3. *In this question you will be assessed on using good English, organising information clearly, and using specialist terms where appropriate.* Describe how to carry out an investigation to test the hypothesis. Include the equipment needed and how to use it, the measurements to make, how to make it a fair test, and a risk assessment.   [9]
4. Use your research to outline another possible method, and explain why you did not choose it.   [3]
5. Draw a table to record data from the investigation. You may use ICT if you wish. [2]

Section 2   Total 30 marks

*Use the **Data Sheet** to answer these questions.*

1. (a) State the independent and dependent variables, and one control variable.   [3]
   (b) The smallest scale division on a measuring instrument is called its resolution. What was the resolution of the instrument you used, and was this resolution suitable for your experiment?   [3]
   (c) Display the **Group A data** on a bar chart or line graph. *This data has been provided for you to use instead of data that you would gather yourself.*   [4]
   (d) Does the **Group A data** support the hypothesis? Explain how.   [3]
   (e) Describe the similarities and differences between the **Group A data** and the **Group B data**. Suggest one reason why the results of the two groups may be different. *The **Group B data** has been provided for you to use instead of data that would be gathered by others in your class.*   [3]
   (f) Suggest how your method might have helped Group A to achieve results that show a clear pattern.   [3]
2. (a) Sketch a bar chart or line graph of the results in **Case study 1**.   [2]
   (b) Explain to what extent the data from **Case studies 1–3** support the hypothesis.   [3]
   (c) Use **Case study 4** to describe the relationship between the position of a Group 1 metal in the periodic table, and the ease of decomposition of its nitrate. Explain how well the data supports your answer. [3]
3. The context of this investigation (the topic it relates to) is determining how to extract metals from ores that contain metal carbonates. Describe how your results may be useful in this context.   [3]

P1 Controlled Assessment-style questions

Hypothesis: It is suggested that the colour of a beaker affects the amount of infrared radiation (IR) it emits. You can investigate this using two copper beakers, one painted matt black, the other painted white. The more IR the beaker emits, the faster it cools.

> Download the Research Notes and Data Sheet for P1 from **www.oxfordsecondary.co.uk/ aqacasestudies**.

Research

*Record your findings in the **Research Notes table**.*

1. Research two different methods that could be used to test the hypothesis.
2. Find out how the results of the investigation might be useful in designing clothing for expeditions to the North Pole.

Section 1   Total 20 marks

Use your research findings to answer these questions.

1. **(a)** Name two sources that you used for your research.
   (b) Which of these sources did you find more useful, and why? [3]
2. **(a)** Identify one control variable.
   (b) Briefly describe a preliminary investigation to find a suitable value for this variable, and explain how the results of this work will help you decide on the best value for it. [3]
3. *In this question you will be assessed on using good English, organising information clearly, and using specialist terms where appropriate.* Describe how to carry out an investigation to test the hypothesis. Include the equipment needed and how to use it, the measurements to make, how to make it a fair test, and a risk assessment. [9]
4. Use your research to outline another possible method, and explain why you did not choose it. [3]
5. Draw a table to record data from the investigation. You may use ICT if you wish. [2]

Section 2   Total 30 marks

*Use the **Data Sheet** to answer these questions.*

1. **(a)** Name the independent and dependent variables, and give one control variable. [3]
   (b) The smallest scale division on a measuring instrument is called its resolution. What was the resolution of the instrument you used, and was this resolution suitable for your experiment? [3]
   (c) Display the **Group A data** in a line graph of temperature decrease against the colour of the beaker. *This data has been provided for you to use instead of data that you would gather yourself.* [4]
   (d) Does the **Group A data** support the hypothesis? Explain how. [3]
   (e) Describe the similarities and differences between the **Group A data** and the **Group B data**. Suggest one reason why the results of the two groups may be different. *The **Group B data** has been provided for you to use instead of data that would be gathered by others in your class.* [3]
   (f) Suggest how your method might have helped Group A identify a clear pattern in their results. [3]
2. **(a)** Sketch a line graph of the results in **Case study 1**. [2]
   (b) Explain to what extent the data from **Case studies 1–3** support the hypothesis. [3]
   (c) Use **Case study 4** to describe the relationship between the surface area of an object and the amount of IR it emits. Explain how well the data supports your answer. [3]
3. The context of this investigation (the topic it relates to) is designing clothing for expeditions to the North Pole. Describe how your results may be useful in this context. [3]

B1 Part 1

Diet, exercise, hormones, genes, and drugs

Why study this unit?

To stay healthy, you need to understand how your body works so you can adopt the behaviours that keep you healthy.

In this unit you will learn what you need to eat and how to exercise to stay healthy. You will find out about infectious diseases, and how your immune system and medicines can deal with them. You will also find out how your hormones play a key role in your growth and development and in helping your body to function properly.

You will learn how drugs affect your health, and what makes us different from each other.

You should remember

1   You are made of cells which are organised into tissues, organs, and systems.

2   You started as one cell which formed stem cells that then became specialised to do different jobs.

3   Your skin and stomach acids try to stop microorganisms from entering the body.

4   Your immune system tries to deal with any microorganisms that do enter the body.

5   Healthy eating, exercise, and using medicines wisely can keep you healthy.

6   Some drugs are harmful.

7   Genes in your body control your characteristics; they are inherited from your parents.

8   Genes are found on your chromosomes.

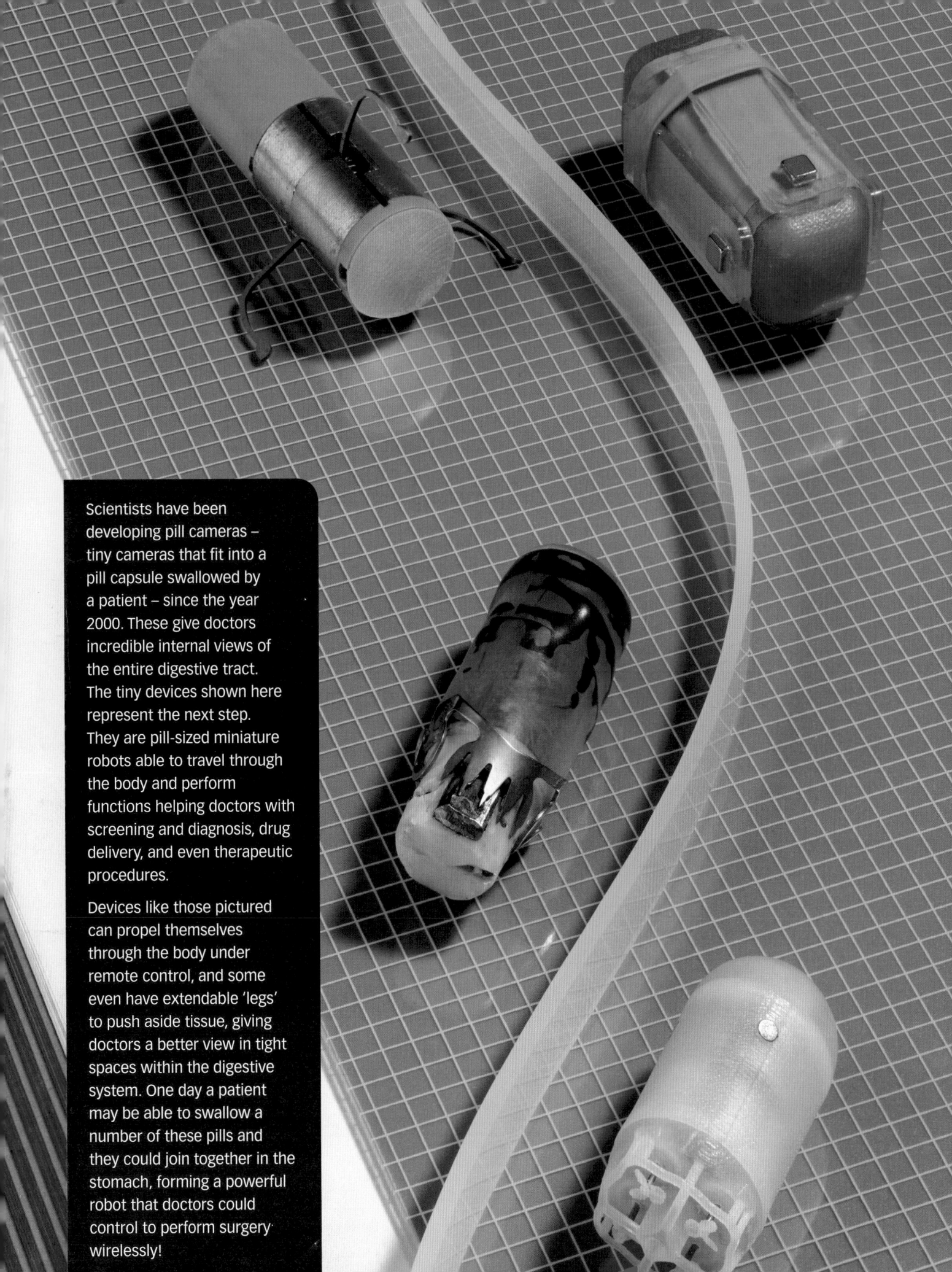

Scientists have been developing pill cameras – tiny cameras that fit into a pill capsule swallowed by a patient – since the year 2000. These give doctors incredible internal views of the entire digestive tract. The tiny devices shown here represent the next step. They are pill-sized miniature robots able to travel through the body and perform functions helping doctors with screening and diagnosis, drug delivery, and even therapeutic procedures.

Devices like those pictured can propel themselves through the body under remote control, and some even have extendable 'legs' to push aside tissue, giving doctors a better view in tight spaces within the digestive system. One day a patient may be able to swallow a number of these pills and they could join together in the stomach, forming a powerful robot that doctors could control to perform surgery wirelessly!

Learning objectives

After studying this topic, you should be able to:

✔ understand the body's metabolic rate and things that affect it

✔ know the benefits of regular exercise

▲ Different ways of taking exercise

▲ School athletics events give children opportunities for exercise and fun

Metabolism and metabolic rate

You are made of lots of cells, and each cell carries out lots of chemical reactions. These reactions keep you alive. They include things like respiration (which releases energy from food) and making proteins.

All of these chemical reactions in cells are collectively called **metabolism**. The rate at which they go on is your **metabolic rate**.

Measuring your oxygen consumption (use) tells you about your metabolic rate. This is because aerobic respiration, which releases the energy you need for metabolism, uses oxygen.

Your metabolic rate varies with the amount of activity you do and the proportion of muscle to fat in your body, and it may also depend on inherited factors.

> **A** What is metabolism?
>
> **B** What is metabolic rate?

Activity and metabolic rate

When you exercise:

- Your muscles contract more to move your limbs.
- Muscles need energy to contract.
- So muscle cells need to respire more to release more energy from glucose.
- To provide this glucose, you need to eat more food.
- Exercise increases the amount of energy expended by the body.

People who exercise regularly and often are rarely overweight. They are usually healthier than people who take little exercise. They do not have the health risks linked to being overweight.

> **C** Why does exercise increase the body's need for energy?

The proportion of muscle to fat in your body

Fat cells store fat. They are inactive and have a low rate of respiration to ensure that they do not use their stored fat.

Muscle cells are active. They need energy from respiration to contract. Their rate of respiration is high.

If you have more muscle and less fat, then your body has a higher metabolic rate than someone of the same size who has less muscle and more fat. You need to eat more food to supply the energy needed for the muscle cells.

Males have more muscle and less fat in their bodies. Females have less muscle and more fat. This is largely determined by genes, although females can increase their muscle mass by exercise.

Most males need to eat more than most females.

> **D** Explain why men need to eat more food than women.

Inherited factors and metabolic rate

- Some people have higher metabolic rates than others.
- Tall people have higher metabolic rates, as they lose more heat from their body surface.
- Contrary to popular belief, overweight people have a higher metabolic rate than slimmer people, as their bodies are larger and need more energy to work.
- People with an underactive thyroid gland have lower metabolic rates.

▲ Woman doing weight training in a gym. This increases her muscle mass.

Key words

metabolism, metabolic rate

Did you know...?

People who exercise use up more food and are usually slim. They are at less risk of heart disease, cancer, diabetes, and arthritis. They also make chemicals in the brain that make them feel good and happy. Exercise also strengthens the immune system so they do not get many infections. However, too much exercise can harm joints and weaken the immune system.

Exam tip

- ✓ You need to mention 'and often' as well as 'regularly' when describing how people should take exercise. After all, once a year is regular!

Questions

1. Describe six ways that regular exercise can improve your health.
2. People under the age of 20 years have higher metabolic rates than those over 20. Why do you think this is?
3. List other inherited factors, not mentioned in the list above, that can affect metabolic rates.

E
↓
C
↓
A*

Learning objectives

After studying this topic, you should be able to:

✔ know what makes a healthy diet
✔ understand that inherited factors also affect health

Key words

balanced diet, deficiency disease, obese, statins

▲ In a healthy, balanced diet the largest part should be carbohydrates. The next layer of the pyramid is fruit and vegetables. These provide many vitamins and minerals and fibre. The next layer is protein foods. Finally, at the top of the pyramid are fats and oils, and sweets, which should be eaten sparingly.

What is a balanced or healthy diet?

A **balanced diet** contains the right amount of different foods and the right amount of energy to keep you healthy.

Food/nutrient	Why you need to eat it
carbohydrates	for energy
fats	for energy
proteins	to build cells and repair tissue
mineral ions and vitamins	needed in small amounts to keep the body functioning healthily

You also need water and fibre.

- Your cells contain about 70% water and you lose it in sweat, breath, tears, faeces, and urine.
- Fibre helps prevent constipation and bowel cancer.

A What is a balanced diet?
B Why do you need to eat carbohydrates, fats, and proteins?
C Why do you need to eat vitamins and minerals?
D Why do you need to drink water?

Malnourishment

Some people do not eat a balanced diet. They eat the wrong amount or wrong types of food. A person is malnourished if their diet is not balanced.

Being malnourished (*mal* means 'bad') may lead to being too fat or too thin. It may also lead to a **deficiency disease**. Rickets is a deficiency disease caused by a lack of vitamin D in the diet. Children with rickets have soft bendy bones that can become deformed. They may have a curved spine or enlarged skull.

Underweight

Some people do not eat enough, or exercise too much. If the energy content of the food you eat is less than the energy you use, you lose body mass. This makes people underweight and can cause health problems. Very thin girls and women have irregular periods.

Overweight

If people eat too much they become overweight. If they are very overweight they are **obese**. Being overweight causes health problems such as type 2 diabetes.

However, good health is influenced by many factors, and diet is just one of them. Inherited factors also play a part. Genes (inherited factors) can affect cholesterol levels.

Cholesterol also affects our health

Cholesterol is a waxy substance. It is only present in animal tissues. You need some because:

- It helps strengthen your cell membranes.
- Your body uses it to make your sex hormones, and to make vitamin D.

However, having too much cholesterol in your blood increases your risk of heart disease.

You get some cholesterol directly from eating animal products like meat, prawns, and eggs. Your liver also makes it from saturated fats that you eat, like butter, chocolate, and fatty meat.

Some people have a lot of cholesterol in their blood because of their genes. Their livers make too much cholesterol. They can take drugs called **statins** to reduce the cholesterol made by the body. They take the statins in the evening because the body makes cholesterol during sleep. Statins inhibit an enzyme involved in making cholesterol.

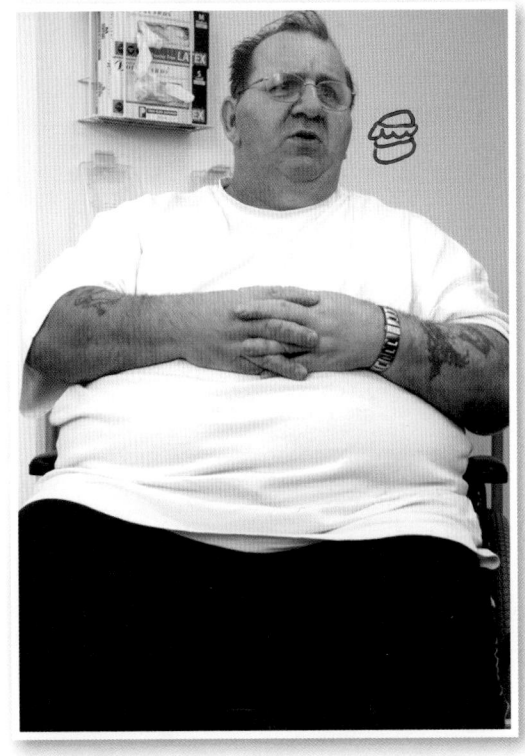
▲ Obesity

Did you know...?

There are one billion obese people in the world. There are also one billion people who do not get enough to eat.

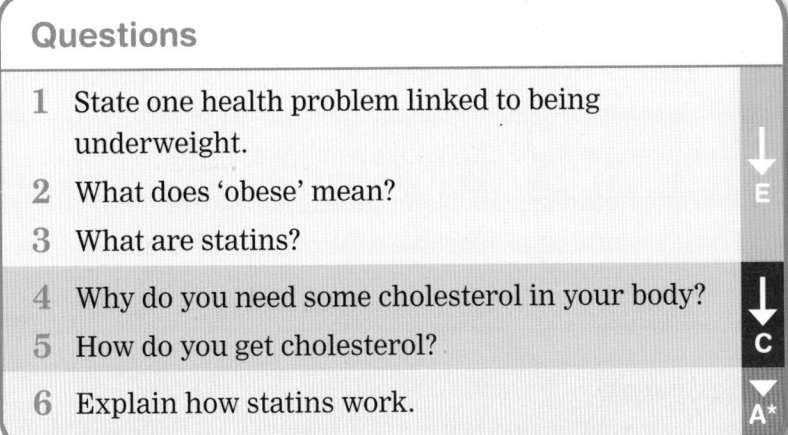

Questions

1. State one health problem linked to being underweight.
2. What does 'obese' mean?
3. What are statins?
4. Why do you need some cholesterol in your body?
5. How do you get cholesterol?
6. Explain how statins work.

↓ E

↓ C

▼ A*

Learning objectives

After studying this topic, you should be able to:

- ✔ state that microorganisms that cause infectious diseases are called pathogens
- ✔ know that bacteria and viruses reproduce rapidly inside the body
- ✔ know that some bacteria produce toxins

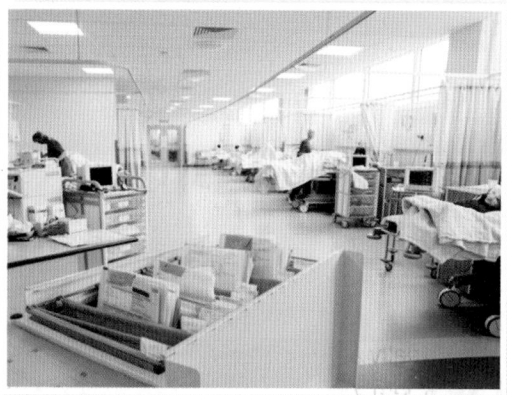

▲ In the mid-twentieth century antibiotics were used to kill bacteria. Unfortunately some bacteria developed resistance to the antibiotics so hospital-acquired infections are now difficult to treat.

▲ Bacteria. Some are rod shaped (shown here as grey), some are round (red), and some are spiral (cream).

Stopping infections spreading

Before the nineteenth century, people did not know about **microorganisms**. Diseases were often passed on from one person to another because doctors did not wash their hands.

A Hungarian doctor called Ignaz Semmelweis was the first to realise that washing hands helps prevent the spread of disease. He noticed that the number of deaths in one hospital maternity ward was higher than in another. At this time nobody knew about germs, but when he investigated he found that on the ward with more deaths the doctors went straight from dissecting bodies to delivering babies. He asked doctors to wash their hands between tasks and the number of deaths reduced significantly.

Hospitals today have codes of practice and strict hygiene regimes for staff, patients and visitors. There are alcohol-based hand gels at the entrance to each ward and by each bed.

- Visitors should clean their hands before entering a ward and when leaving a ward.
- Nurses should clean their hands after dealing with one patient and before going to the next patient.
- Patients should also regularly clean their hands.

Hands spread many microorganisms because we touch:

- food
- door knobs
- surfaces
- each other.

However, some people are not as conscientious as they should be about using the hand gels.

> **A** Why should visitors clean their hands before entering a ward and when leaving a ward?
>
> **B** Why should nurses clean their hands after dealing with a patient and before dealing with the next patient?

Pathogens

Any microorganism that can cause an infectious disease is called a **pathogen**. Some **bacteria** are pathogens. All **viruses** are pathogens.

Bacteria

Not all bacteria are pathogens. We have millions of them on our skin and in our gut and we could not live without them. However, if they get through our gut wall or skin and into our blood or cells they can make us very ill. They reproduce rapidly inside our warm bodies and some produce **toxins** (poisons) that make us feel ill. Some bacterial toxins may cause death.

Viruses

It is debatable whether viruses are living or not. They are not made of cells. They cannot carry out any life processes. They have to insert themselves into a host cell and hijack that cell's parts to make copies of the virus. These new virus particles can then burst out of the cell and infect many other host cells in a 'chain reaction'. Viruses damage and destroy our cells when they infect us in order to reproduce. Viruses are much smaller than bacteria.

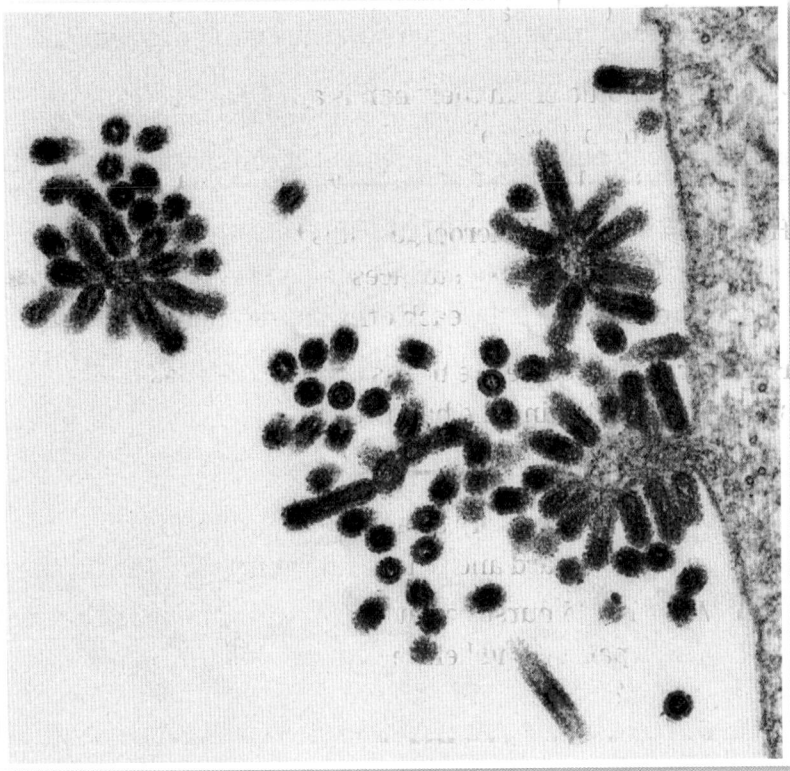

▲ Coloured transmission electron micrograph (TEM) of influenza viruses, shown in red, that have reproduced inside a cell of the respiratory tract and are breaking out of the cell. They have damaged the host cell. (× 150 000).

Key words

microorganism, pathogen, bacteria, virus, toxin

Did you know...?

Although all viruses are pathogens, they don't all infect humans. Some infect animals or plants and some live in the sea and infect algae. This is useful to the environment as it keeps the algal growth in check. Some viruses attack bacteria and enter their cells.

Exam tip   AQA

✔ Remember that viruses cannot produce toxins. They are not really alive and do not have any metabolic processes or structures for making proteins.

Questions

1 What are pathogens?
2 Do viruses infect other species besides humans?
3 How do viruses damage our cells?
4 Why is it hard to say whether viruses are truly living?

A Why is pain useful?

B What are the symptoms of many infectious diseases?

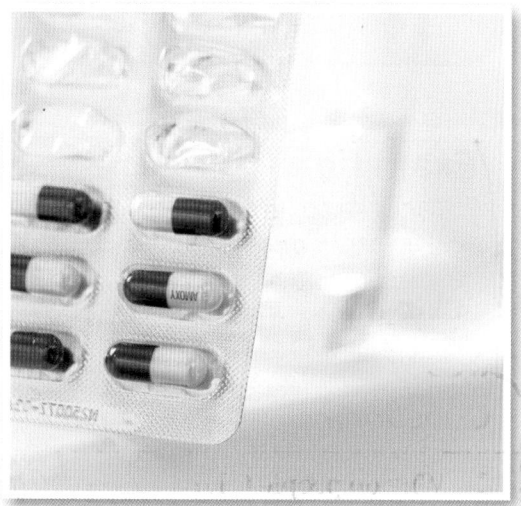

▲ Amoxicillin antibiotic pills. Amoxicillin is a type of penicillin.

C What are antibiotics?

D Name the process by which strains of bacteria develop resistance to antibiotics.

How resistance to an antibiotic develops ▶ in bacteria. A new strain of bacteria that is resistant to the antibiotic has developed. Some of these bacteria may infect other people and, when these people are given the antibiotic, it does not work.

Why do you take painkillers when you have an infection?

Pain is useful because it tells you that something is wrong. Many infections give us a headache and aching muscles. We may also get a fever (high temperature) or feel shivery. However, we all want to carry on with our lives while we get better, so relieving the painful **symptoms** is useful. **Painkillers** such as paracetamol or codeine can do this. They do not kill the pathogens but make us feel better while our immune system or antibiotics (or both) kill the pathogens.

Antibiotics

Antibiotics have been available since the 1940s and have saved millions of lives. Antibiotics are medicines that can kill infective bacteria inside the body. They can help cure diseases caused by bacteria, such as chest or ear infections, TB, or blood poisoning. Penicillin is an antibiotic. It does not kill all types of bacteria. Since it was discovered, scientists have found many more types of antibiotics. Specific antibiotics target specific bacteria. Using the appropriate antibiotic can help reduce the spread of antibiotic resistance.

Resistance to antibiotics

Many strains of bacteria have developed **resistance** to various antibiotics. This resistance has developed as a result of chance (spontaneous) mutation and **natural selection**.

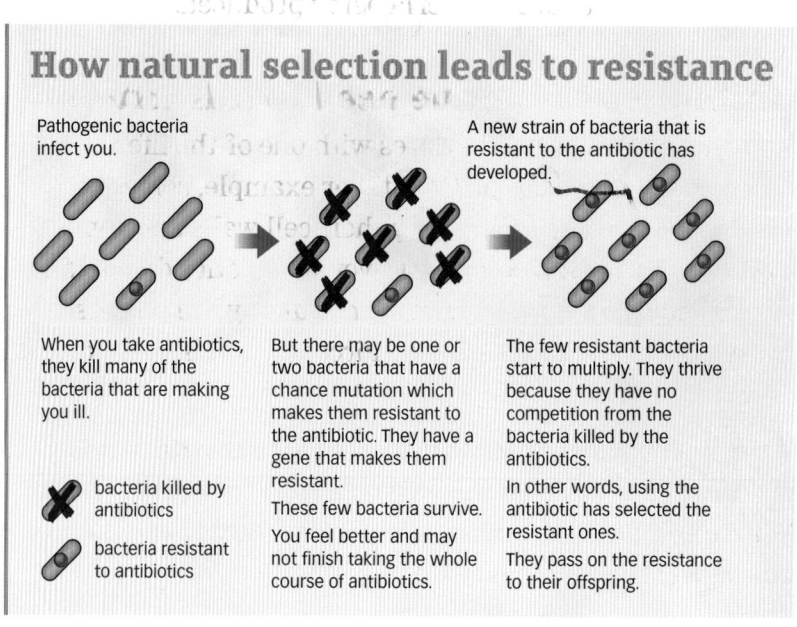

How natural selection leads to resistance

Pathogenic bacteria infect you.

A new strain of bacteria that is resistant to the antibiotic has developed.

When you take antibiotics, they kill many of the bacteria that are making you ill.

✖ bacteria killed by antibiotics

🦠 bacteria resistant to antibiotics

But there may be one or two bacteria that have a chance mutation which makes them resistant to the antibiotic. They have a gene that makes them resistant.

These few bacteria survive.

You feel better and may not finish taking the whole course of antibiotics.

The few resistant bacteria start to multiply. They thrive because they have no competition from the bacteria killed by the antibiotics.

In other words, using the antibiotic has selected the resistant ones.

They pass on the resistance to their offspring.

MRSA

One of the most troublesome strains of bacteria that has developed resistance to antibiotics is MRSA. Its full name is methicillin-resistant *Staphylococcus aureus*. Methicillin is a very strong antibiotic only used in bad cases of infection, but this bacterium is resistant to even that. It causes many hospital-acquired infections each year and many deaths.

This bacterium lives on our skin and in our noses where it does no harm. But if your skin is not clean when you have an operation and the bacterium gets into a wound, it can cause a dangerous infection which could kill you. However, this bacterium can easily be killed with antiseptics and disinfectants, so it is important that hospitals practise good hygiene.

How to prevent further resistance arising

In the 1940s and 1950s, when antibiotics were new, many doctors prescribed them when they were not really needed. Patients did not always take the whole course of the drugs, but stopped when they began to feel better. This overuse of antibiotics increased the rate at which antibiotic resistance developed.

Nowadays, doctors are aware of the dangers of overusing antibiotics. They do not prescribe them for non-serious infections like sore throats. This is to slow down the rate of resistant strains of bacteria being produced.

Antibiotics cannot be used to kill viruses

Each type of antibiotic interferes with one of the life processes that bacteria carry out. For example, some antibiotics stop bacteria making their cell walls and some stop them making proteins. However, antibiotics do not kill viruses because viruses do not carry out any life processes. Viruses get inside your cells and force your cells to make more copies of them.

It is difficult to make a drug that kills viruses without also damaging your cells, because viruses have to be inside your cells to reproduce. There are drugs called antivirals that stop your cells making copies of viruses.

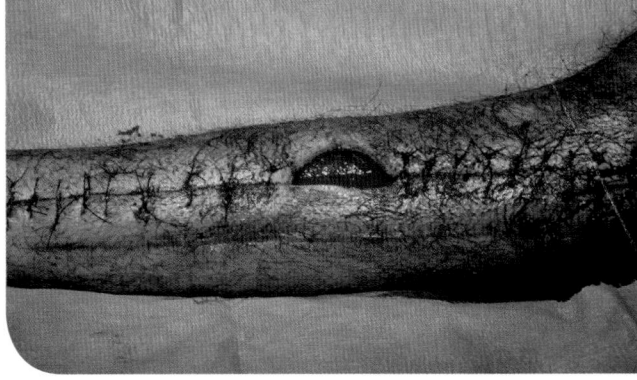

▲ Wound infected with MRSA. The spread of MRSA can be prevented by doctors and nurses washing their hands between treating patients.

Key words

symptoms, painkiller, antibiotic, resistance, natural selection

Exam tip  AQA

✔ Remember that it is the bacteria that can become resistant to antibiotics, not the patient.

Questions

1 Why do people take painkillers when they have an infection?

2 Why is MRSA such a problem?

3 How can doctors and patients prevent further antibiotic resistance happening?

4 Explain why antibiotics cannot be used to kill viruses.

5 Why is it difficult to make drugs which cure viral infections?

Learning objectives

After studying this topic, you should be able to:

✔ describe how the immune system deals with pathogens

✔ know that people can be immunised against some diseases

Key words

immune system, antibody, antigen, immunity, immunisation, epidemic, pandemic

A How does your body stop pathogens entering it?

Did you know...?

Your immune system also protects you from cancer. It can recognise and kill cancerous cells.

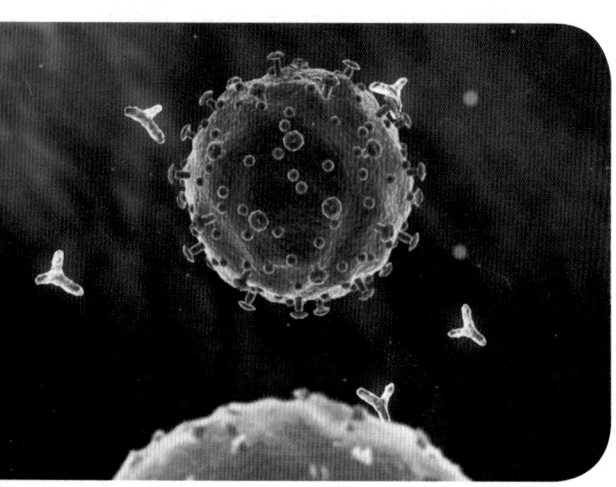

▲ Antibodies surrounding a virus particle

How your immune system deals with pathogens

Your body has barriers to stop pathogens entering it. These include

- skin
- tears
- blood clotting when you cut yourself.
- stomach acid
- mucus in the airways

However, sometimes some pathogens manage to enter your body. When they do, your **immune system** may deal with them. This involves your white blood cells.

White blood cells

You have different sorts of white blood cells.

- Phagocytes engulf (ingest) the pathogens.
- Lymphocytes produce antibodies or antitoxins.

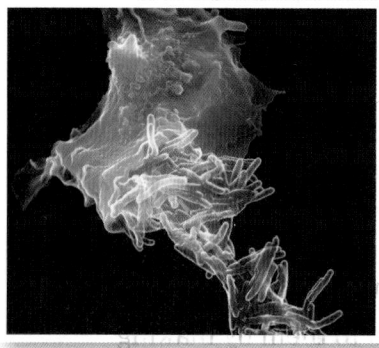
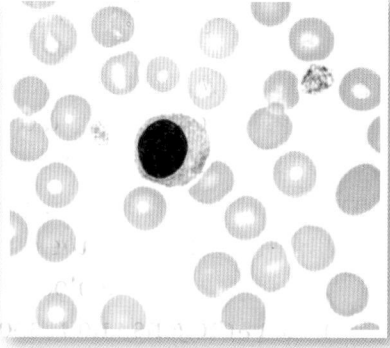

▲ Coloured scanning electron micrograph (SEM) of a phagocyte ingesting TB bacteria (× 3500)

▲ Light micrograph of blood showing lymphocyte (stained purple) and red blood cells. Lymphocytes are white blood cells that make antibodies (× 400).

Antibodies

Antibodies are proteins. Each type of antibody can destroy a particular type of bacterium or virus. This is because:

- Each type of pathogen has particular **antigens** (proteins) with a specific shape on its surface.
- Each type of antibody (also a protein) has a particular shape and can lock on to a particular antigen.
- Your immune system makes the right sort of antibodies to lock on to the antigens of the particular pathogen that is in your body.
- Once the pathogen is coated with antibodies, white blood cells can ingest and kill the pathogens.

Once you have recovered from an infection, you have **immunity** to it. If that same pathogen enters your body again, your body makes antibodies so quickly that the pathogen is destroyed before it makes you feel ill.

> B Describe two ways that white blood cells can kill pathogens.
>
> C Explain how you become immune to a disease such as measles.

Immunisation

Many people used to die from infections like TB and smallpox. In the developing world today many die from measles, malaria, and cholera. **Immunisation** can make people immune to a disease, without them having the disease. Widespread immunity in a population reduces the spread of infection within it.

Here's how it works:

- A small amount of dead or inactivated pathogen is introduced into your body – usually by injection. This is called being vaccinated. The dead or inactivated pathogens still have the antigens on their surface.
- Some of your white blood cells recognise these antigens on the pathogens and respond to them by making antibodies.
- If, later on, the live pathogens get into your body, your white cells quickly make the right sort of antibodies.
- These antibodies destroy the pathogens before they can make you ill.

How do mutations affect vaccines?

Some viruses, like the flu virus, mutate often. This causes them to have slightly different antigens. Your immune system does not recognise these viruses, so they can make you ill again with flu, even though you may have had flu before. So every year new vaccines are made for the new strains of flu that are likely to infect people that year.

In 2009 many people in many countries were immunised against swine flu to prevent the disease sweeping across countries and causing an **epidemic**, or across continents and causing a **pandemic**.

▲ This baby is having an MMR vaccination to protect her against measles, mumps, and rubella

> D Draw a flow diagram to show how immunisation works.

Exam tip $\quad$ AQA

- ✔ Do not confuse the terms 'immunity' and 'resistance'. People become immune to infectious diseases because they have an immune system. Bacteria (not people) may be resistant to antibiotics.

Questions

1 Describe one way that pathogens can enter your body. $\quad$ ↓ E

2 What is (a) an antigen, and (b) an antibody?

3 Describe two ways that you can become immune to mumps. $\quad$ ↓ C

4 Explain how mutations in viruses can lead to epidemics.

5 Healthcare workers are vaccinated each year against flu. Why do you think this is? $\quad$ ↓ A*

Learning objectives

After studying this topic, you should be able to:

✔ understand the role and organisation of the nervous system
✔ know that receptors detect stimuli
✔ know that nerve impulses pass from receptors along neurones
✔ recall how reflex actions come about

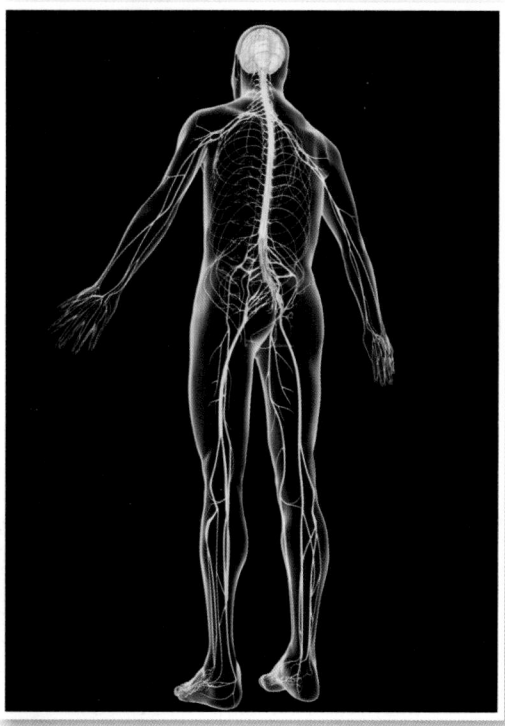

▲ The nervous system is made up of the central nervous system (CNS) and the peripheral nervous system.

Key words

stimulus, central nervous system, peripheral nervous system, receptor, neurone, reflex action, synapse, effector

Why you need to respond to change

You, and all living things, need to be able to respond to changes in the environment. These changes are called **stimuli**. If you could not detect and respond to stimuli you would not be able to find food or avoid danger. You also need to learn from your experiences and coordinate your behaviour.

The structure of the nervous system

There are two main parts to the nervous system:

* The **central nervous system** (CNS) – the brain and spinal cord.
* The **peripheral nervous system** – nerves taking information from sense organs into the CNS, and nerves taking information from the CNS to effectors (muscle or glands).

A What is a stimulus?
B Why do we need to be able to detect stimuli?
C List four stimuli that your skin can detect.

Sense organs or receptors

Receptor cells are special cells adapted to detect stimuli. Like most animal cells they have

* a nucleus
* a cell membrane
* cytoplasm.

Information from the receptors passes as electrical impulses. It travels along nerve cells called **neurones** to the brain. The brain then coordinates the response. Some responses are voluntary – they are consciously controlled by the brain. For example you may hear part of a song on the radio and decide whether to listen and turn up the volume or to switch off the radio.

Reflex actions

Sometimes you need to respond to a potentially dangerous situation very quickly. For example, if you touch a hot object, you need to quickly withdraw your hand before it burns.

There is no time to think, so the brain does not need to be involved. Instead the response is coordinated by the other part of the CNS, the spinal cord. These responses are called **reflex actions**. They are fast, automatic, and protective.

Each reflex action follows the pathway: stimulus → receptor → sensory neurone → relay neurone → motor neurone → synapses → effector → response.

This pathway is described as a reflex arc.

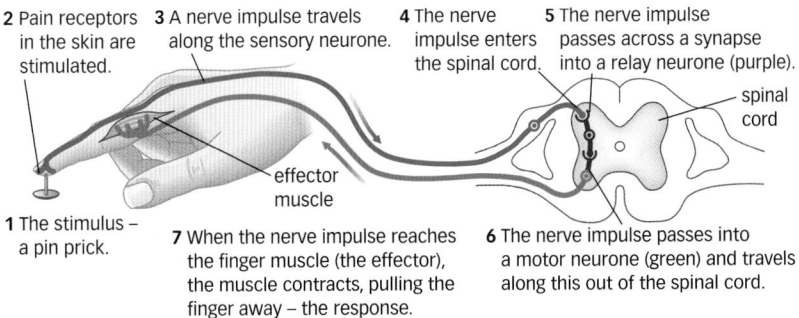

2 Pain receptors in the skin are stimulated. **3** A nerve impulse travels along the sensory neurone. **4** The nerve impulse enters the spinal cord. **5** The nerve impulse passes across a synapse into a relay neurone (purple).

spinal cord

effector muscle

1 The stimulus – a pin prick.

7 When the nerve impulse reaches the finger muscle (the effector), the muscle contracts, pulling the finger away – the response.

6 The nerve impulse passes into a motor neurone (green) and travels along this out of the spinal cord.

▲ A reflex arc. The impulse goes from receptor to CNS and then to effector to bring about the response. The relay neurone inside the spinal cord coordinates the response by connecting the sensory neurone to an appropriate motor neurone. The information travels from one neurone to another across a small gap called a **synapse**.

Effectors
Effectors are glands or muscles that carry out a response.
- A muscle responds by contracting.
- A gland responds by secreting chemical substances.

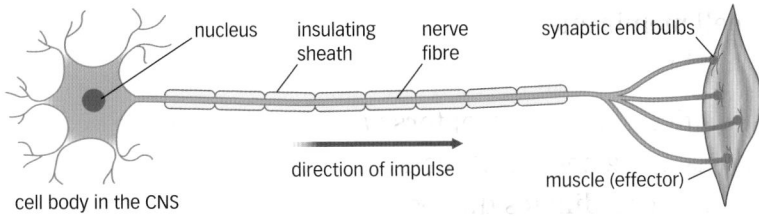

nucleus   insulating sheath   nerve fibre   synaptic end bulbs

direction of impulse   muscle (effector)

cell body in the CNS

▲ Structure of a motor neurone. The nerve impulse is carried along the nerve fibre.

Neurones are adapted to their functions because they are long, have an insulating sheath to prevent impulses leaking away, and have branched endings so they can communicate with many other neurons.

Examples of reflex actions include:
- the knee-jerk reflex
- withdrawing a hand from a hot plate.

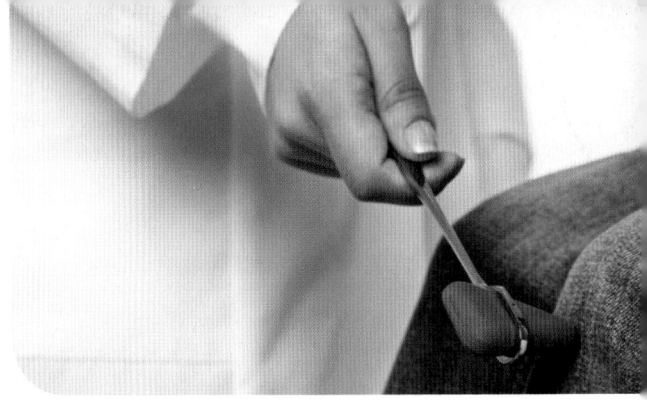

▲ The knee-jerk reflex test. When the leg is tapped just below the knee, the leg straightens. This reflex is used when we walk.

Did you know...?

The pupil-shrinking reflex does involve the brain, but not at the conscious level, so it is still very quick and automatic.

Exam tip   AQA

✔ When you are writing about nerve transmission, always use the technical term 'impulses'. Don't write about 'signals' or 'messages' or you will lose marks.

Questions

1 What are voluntary responses?

2 Why are reflexes automatic and rapid?

3 How does the pupil-shrinking reflex protect the eye?

4 Which part of the neurone does the impulse travel along?

5 What is the function of the insulating sheath?

↓ E

↓ C

↓ A*

Learning objectives

After studying this topic, you should be able to:

- ✔ know that your internal conditions are controlled
- ✔ know that many processes in the body are coordinated by hormones

Key words

hormone, secrete, gland, target organ, ion

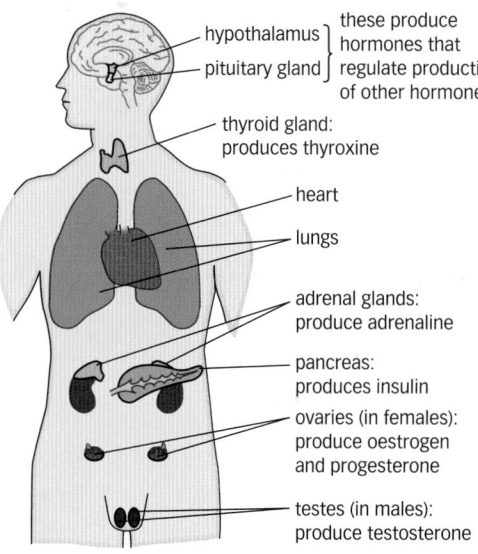

hypothalamus ⎫
pituitary gland ⎭ these produce hormones that regulate production of other hormones

thyroid gland: produces thyroxine

heart

lungs

adrenal glands: produce adrenaline

pancreas: produces insulin

ovaries (in females): produce oestrogen and progesterone

testes (in males): produce testosterone

▲ Some of the main glands and the hormones they produce

A Write down four internal conditions that have to be controlled.

B How does your body lose water?

Internal conditions are controlled

Many things inside your body are controlled. Their levels are kept within a narrow acceptable range. In this section you will learn about

- the water content of the body
- the ion content of the body
- body temperature
- blood sugar levels.

Hormones

Hormones are chemicals. They

- regulate the functions of many organs and cells
- coordinate many processes in the body
- are produced (**secreted**) from **glands** into the bloodstream
- travel in the blood to **target organs**.

The body usually reacts slowly to hormones. Hormones coordinate long-term body changes such as maturing (growing up).

However, some hormones act quickly. One hormone that acts quickly is made in part of the brain and acts on the kidneys to regulate the water content of the blood.

How the body loses and gains water

You, like all living things, contain about 70% water in your cells. Your cells need this water, otherwise their chemical reactions cannot go on properly. Although your skin is waterproof, you lose water from the:

- skin in sweat
- lungs by breathing out
- kidneys in urine.

The amount of water in your body has to be balanced. You gain water from:

- drinks containing water
- food that contains water
- respiration of digested food.

Your kidneys help regulate the water content of your body. A hormone, made in part of the brain, helps the kidneys to only pass out water when you have too much.

Blood sugar level

When you eat and digest food, the products of digestion pass into your bloodstream. They are then carried to cells. Sugar is delivered to cells so they can get a constant supply of energy. Your blood sugar level has to be controlled so there is never too much or too little. Hormones regulate the blood sugar level.

The ion content of the body

Your blood and the watery fluid in between all your cells contain **ions** (sometimes called electrolytes). These include sodium ions, potassium ions, magnesium ions, calcium ions, and hydrogen ions. They are all very important in helping your nerves to work and keeping the body fluids at the right pH. Hormones work with the kidneys to make sure the balance of these ions is right. You may pass out unwanted ions in urine.

Temperature

Your body temperature has to be maintained at around 37 °C. This is warm enough for the chemical reactions in cells to happen quickly enough to keep you alive. The enzymes that control these chemical reactions work well at this temperature. If your body temperature was much higher, some enzymes and many of the other proteins in your body would stop working.

The heat made during respiration keeps your body warm. When you exercise, your muscle cells are working harder and need to respire more. This makes more heat. Your body gets rid of the heat by sweating. You also lose heat when you breathe out warm air, and you lose it from the warm blood flowing near the skin surface.

Did you know...?

Some animals, like desert rats, are well adapted to living in dry places. They do not drink but get all of their water from food. They do not sweat but keep cool by finding shade.

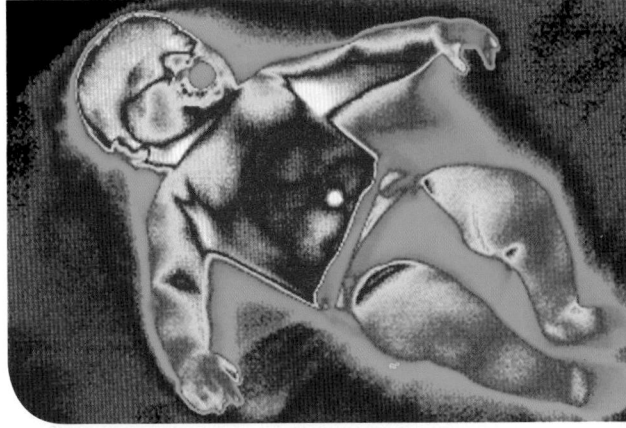

▲ Thermal image of a baby. The hottest parts look white. Then the scale goes from red, yellow, green, blue, to purple (coolest).

Exam tip — AQA

✔ If you are asked what a hormone is, give as much information as possible. Don't just say it is a chemical. Say where hormones are made (glands) and that they travel in the blood to target organs, to coordinate body processes.

Questions

1 What are hormones?

2 Name four ions in your body fluids.

3 Why do you need to have the right amounts of ions in your body fluids?

4 How does your body get rid of unwanted (excess) ions?

5 How does your body (a) generate heat; (b) lose heat?

6 Explain why your body temperature needs to be maintained within narrow limits at around 37 °C.

ประกอบ

Learning objectives

After studying this topic, you should be able to:

✔ know that several hormones are involved in the menstrual cycle

✔ describe the role of FSH and oestrogen in the menstrual cycle

✔ explain how oestrogen and progesterone can be used as contraceptives

Did you know...?

At puberty males produce small amounts of oestrogen and progesterone and females produce small amounts of the male sex hormone, testosterone. In females testosterone causes pubic hair and hair under the arms to grow.

A Where is FSH made in the body?

B What does FSH do?

C When during the menstrual cycle is the level of FSH highest?

Growing up

At puberty, children's bodies begin to change into those of sexually mature adults. This change takes several years. The first thing that happens is that the ovaries in females and testes in males develop and begin to produce the **sex hormones**.

In females the sex hormones **oestrogen** and **progesterone** are made in the ovaries. These hormones are involved in the **menstrual cycle**.

The menstrual cycle

At puberty females begin to have a menstrual period each month. Several hormones help coordinate the menstrual cycle:

- The pituitary gland in the brain produces a hormone called follicle stimulating hormone (**FSH**).
- FSH causes eggs in the ovaries to mature, one each month.
- It also stimulates the ovaries to make oestrogen.
- Oestrogen stimulates the pituitary gland to make luteinising hormone (**LH**), which triggers the release of an egg (**ovulation**) from the ovary.
- Oestrogen also prevents more FSH being secreted, and it repairs the uterus lining after menstruation.
- Progesterone maintains the uterus lining, and works with oestrogen to prevent secretion of FSH.

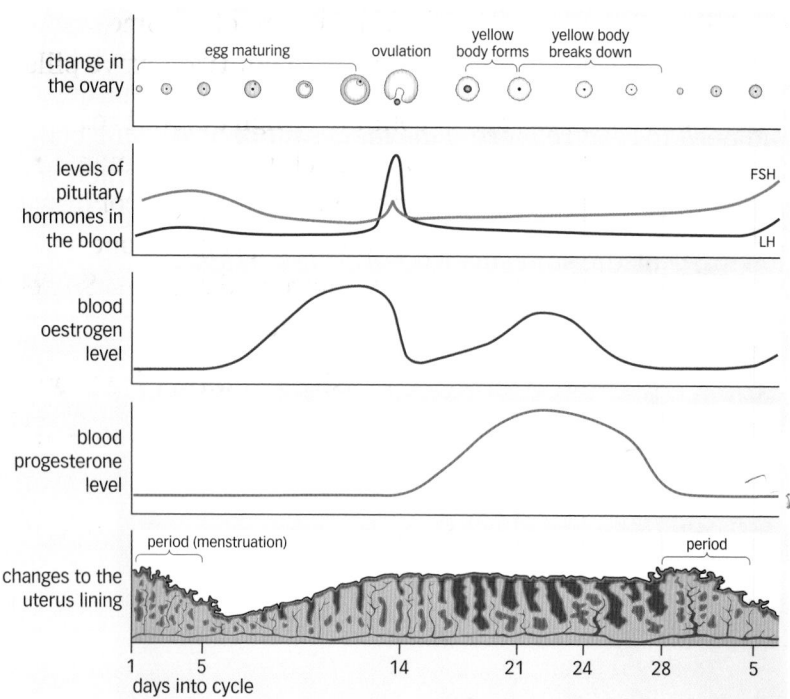

Events in the menstrual cycle ▶

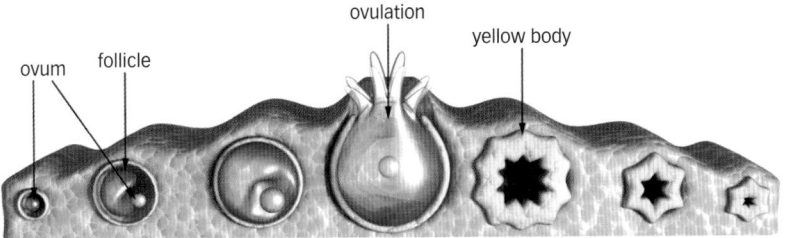

ovulation

ovum   follicle   yellow body

▲ Changes in the ovary during the menstrual cycle. The egg develops in a follicle. It then bursts out of the follicle. The empty follicle develops into a yellow body which makes a hormone, progesterone. This stops menstruation.

If the egg is not fertilised, at the end of the cycle the uterus lining passes out of the body. This is the period. If the egg is fertilised then the uterus lining stays so that the baby can develop.

How the female sex hormones control fertility

The female sex hormones can be used to make a woman less fertile, if she does not want to become pregnant. ท้อง, คุมกำเนิด

During pregnancy both oestrogen and progesterone levels are high and they inhibit FSH production from the pituitary gland. This prevents the development and release of any more eggs.

Scientists realised that if women took these hormones in a daily pill, the high levels in the body would prevent ovulation. Without ovulation women cannot become pregnant. These hormones are used in **contraceptive** pills.

The first birth-control pills contained high amounts of oestrogen. They prevented ovulation but many women suffered from side-effects. Birth-control pills now contain a much lower dose of oestrogen and some progesterone to inhibit egg production. These give fewer side-effects. Some birth-control pills contain only progesterone.

Some women may not conceive easily and want to improve their fertility. FSH may be used to help them conceive. ตั้งครรภ์

Key words

sex hormones, oestrogen, progesterone, menstrual cycle, FSH, LH, ovulation, contraceptive

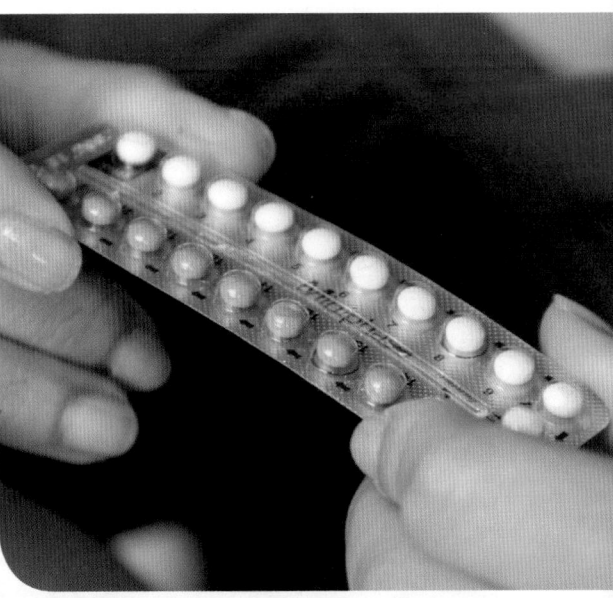

▲ Each pack contains enough contraceptive pills for one month. They are usually taken for 21 days of each month and then not taken for 7 days, so the woman has a period.

Questions

1   State two functions of oestrogen in controlling the menstrual cycle.

2   What is the function of progesterone?

3   During which part of the menstrual cycle is a woman most likely to conceive?

4   Explain how the oestrogen in contraceptive pills prevents pregnancy.

↓ E

↓ C

↓ A*

9: Using hormones to control fertility

Learning objectives

After studying this topic, you should be able to:

- ✔ know that hormones can be used in in vitro fertilisation (IVF) to control fertility
- ✔ evaluate the benefits and problems of using hormones to control fertility

▲ Robert Edwards, British IVF pioneer

Key words

in vitro fertilisation

Test tube baby

You have seen how female hormones are used in contraceptive pills. They have also been used to help infertile women have babies.

In 1978 in the UK, the first 'test tube baby' in the world was born. She is Louise Brown. Her parents, John and Lesley, had been trying to conceive a baby for nine years, but Lesley had blocked oviducts. The procedure that helped them conceive was developed by two British doctors, gynaecologist Patrick Steptoe and physiologist Robert Edwards.

Louise developed inside her mother's womb, not in a test tube. However, the egg was fertilised in a glass dish. The Latin word for glass is '*vitro*' so this is where the term **in vitro fertilisation** (IVF) comes from.

How IVF works

Doctors collect eggs from the woman. Before they do that they inject her with FSH and LH. This causes her ovaries to make more than the usual one egg during her menstrual cycle. The eggs are then collected from the mother and fertilised by mixing them with the father's sperm in a glass dish. To make the procedure more likely to work, sperms are selected and one is injected into each egg. This is done under a powerful microscope.

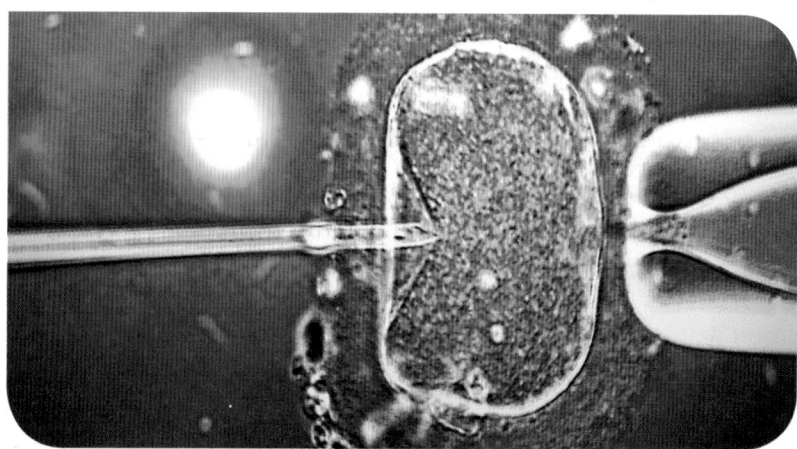
▲ Human sperm being injected into a human egg

The fertilised eggs begin to develop into embryos. When they are tiny balls of cells, two are chosen and inserted into the mother's womb.

Benefits and problems of using hormones to control fertility

Benefits	Problems
• Couples who are infertile can be treated and have their own children.	• Some religions feel that it is wrong for humans to exercise control over their fertility.
• The embryo can be tested for any genetic disorders before being implanted.	• Some people feel IVF is wrong, as some of the embryos made are not allowed to develop into people.
• People can choose when to have children.	• Some people think that infertile couples should adopt instead, as there are many orphaned children without homes.
• They may wish to delay having children whilst developing their career.	
• They may want to wait until they can better afford to look after the children.	• IVF is expensive and is not always successful.
• People can limit the size of their family.	• People may delay having their families for too long and then find that it is difficult to conceive.
• Some people may want to choose not to have children.	• Some women may suffer side-effects of taking contraceptive pills, such as mood swings, weight gain, or feeling sick.

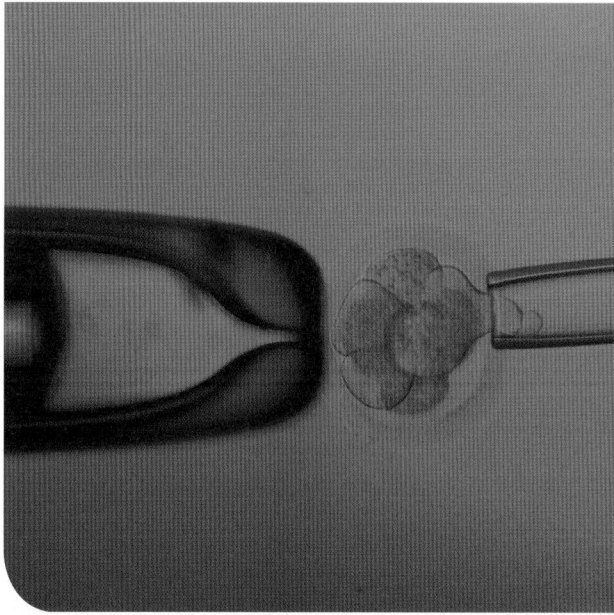

▲ A human embryo produced by IVF, before being implanted into the mother's uterus

Questions

1   Describe how FSH and LH can be used to increase fertility in women who cannot conceive.

2   Why do you think two embryos are implanted into the mother's uterus?

↓ E

3   Evaluate the advantages and disadvantages of IVF.

4   Some other fertility drugs contain a chemical that inhibits oestrogen. Suggest how this can increase fertility.

↓ C

5   Some families have a history of a particular genetic disorder. In these cases, fertilisation can be in vitro and the embryos can be tested for genetic defects. Only healthy embryos will be implanted. Do you think this is a good or bad idea? Give reasons for your answer.

↓ A*

Exam tip AQA

✓ If you are using abbreviations, like IVF, give the full name first, with the abbreviation after it in brackets. Then use the abbreviation.

Learning objectives

After studying this topic, you should be able to:

✔ know that plants are sensitive to light, moisture, and the force of gravity

✔ understand that plants produce hormones to control and coordinate growth

✔ recall how plant hormones can be used in agriculture

▲ Seedlings growing towards light. They are showing a positive phototropic response.

Key words

tropism, phototropic, auxin, geotropic

A What do plant growth hormones control in plants?

B What is (a) phototropism, and (b) geotropism?

Plants respond to their environment

Plants as well as animals respond to stimuli, namely changes in their environment. Plants make chemicals called plant hormones (plant growth substances) that control and coordinate

- the growth of shoots and roots
- flowering
- ripening of fruits.

You may have noticed that plants growing on a windowsill tend to bend towards the light source. If you want them to grow straight you have to keep turning them. A plant's response to a stimulus is called a **tropism**.

Phototropism

Plant shoots grow towards light. They are positively **phototropic**.

- Plant shoot tips are a growing point. They make a plant hormone called **auxin** and this moves down to other parts of the stem.
- When light strikes one side of the shoot tip, more auxin builds up on the *other* side of the shoot tip (furthest from the light).
- This causes the shoot to bend over, towards the light.
- This is useful to the plant, as it needs light to make food.

How does auxin make the shoot bend?

Auxin is unevenly distributed in the shoot tip. This auxin then moves down the stem and causes cells on the side of the shoot furthest from the light to elongate more than those nearest the light.

Geotropism

Geotropism is the response of a plant to gravity. Roots grow downwards in response to the pull of gravity. Auxins may be involved in this response, but other chemicals that inhibit growth may also play a part.

Plant hormones in agriculture and horticulture

Plant hormones can be applied to plants to either speed up or slow down their growth.

▲ Venus flytrap

▲ This sensitive plant's leaves close up after the plant has been touched

Did you know...?

Plants also respond to other stimuli. Plant roots grow towards water.

Some plants respond to touch. Climbing plants put out tendrils and cling to walls or canes.

Some plants move their leaves downwards very quickly when touched. This startles animals trying to eat them and protects the plant. The Venus flytrap responds to flies touching hairs on its leaves. It snaps the leaves shut, trapping the flies so it can digest them.

Weedkillers

Auxins are used as selective weedkillers. Agent Orange was used in the Vietnam War. It made trees lose their leaves. Without leaves, the trees cannot make food and they die.

Rooting powder

If cuttings are dipped into auxin powder, this helps the cuttings make new roots. With new roots the cuttings can anchor in the soil and take up water and minerals.

Fruit ripening

Auxin can be sprayed on to fruit trees to prevent the ripe fruit from dropping. Then it can all be harvested at the same time and the fruit is not bruised by falling to the ground. If a high dose of auxin is sprayed later, it makes all the fruits fall.

Control of dormancy

Seeds are usually dormant when they are shed from the parent plant. This stops them germinating at the end of the summer, as new plants may not survive the harsh winter. However, if commercial growers want to germinate seeds in greenhouses during winter, applying auxin can break the dormancy.

Exam tip ▸  AQA

✔ Remember that although plant growth substances like auxin are called plant hormones, they do not work in the same way as animal hormones.

Questions

1. Why is it an advantage to a plant if the shoots are positively phototropic?
2. Why is it an advantage to a plant if the roots are positively geotropic?
3. Describe four commercial uses of auxins.
4. Explain how auxin causes shoot tips to bend towards the direction of the light source.

▲ Horticulturist dipping a cutting in rooting powder

Learning objectives

After studying this topic, you should be able to:

✔ know about different types of drugs

✔ evaluate why some people use illegal drugs for recreation

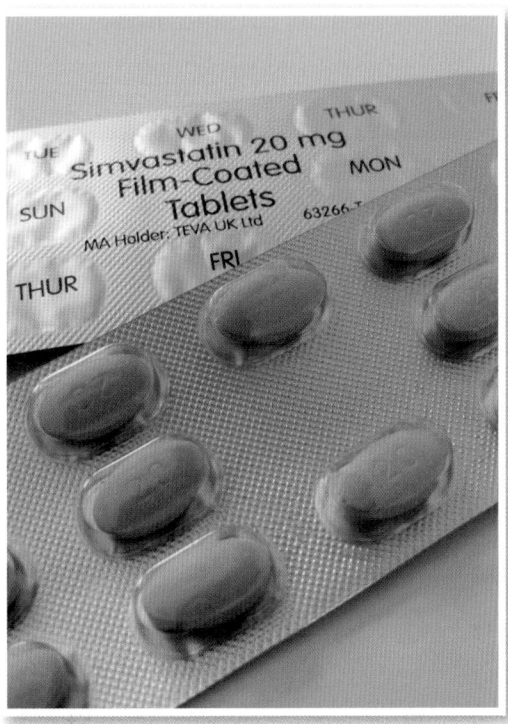

▲ Statins lower blood cholesterol. Some statins cause muscle pain in some people, and a doctor will need to find the best type for each patient.

A Why do you need to have a prescription from a doctor to get certain drugs, like strong painkillers?

B Why do legal drugs have a greater impact overall on people's health than illegal drugs?

Drugs may be beneficial or harmful

A **drug** is a chemical that alters the way your body or brain works. Drugs may alter your behaviour as well as altering your metabolism.

Beneficial drugs are medicines like painkillers, antibiotics, and statins. Some drugs have to be prescribed by a doctor. This is because they may

- have side-effects
- interfere with another medicine the patient is taking
- be harmful for a particular patient if they have another condition
- be harmful if taken too often.

Using drugs for recreation

Legal recreational drugs

Some drugs are legal and used for recreation. These include caffeine, nicotine in tobacco, and alcohol. Caffeine is usually not harmful. Nicotine makes people **addicted** to tobacco, and that causes cancer. Alcohol can harm the nervous system. It alters people's behaviour and may lead to violence or accidents.

Illegal recreational drugs

Some athletes use performance-enhancing drugs like anabolic steroids. These can have harmful side-effects. It may be unethical to use them as it gives some athletes an unfair advantage. The athletes may also suffer side-effects from taking the anabolic steroids.

Progression to hard drugs

Many young people experiment briefly with some types of drugs. Unfortunately some of them may go on to take hard drugs like heroin and cocaine. Both these drugs are very addictive. When users try to stop taking them they get **withdrawal symptoms**.

Impact on health

All drugs have an effect on your health. However, the legal drugs, like alcohol and tobacco, have a greater overall impact and cause more harm. This is because more people use them so more people are harmed.

▲ Field of opium poppies in Dorset. Opium is obtained from the seed heads of this poppy. Opium contains morphine and codeine. Opium can also be refined to make the illegal drug heroin.

Cannabis

Some people believe cannabis is a very good painkiller. People with multiple sclerosis find it relieves their symptoms. However, some people have concerns that the chemicals in cannabis smoke may

- lead to mental health problems in some people
- lead the user on to addiction to hard drugs like heroin and cocaine
- increase the risk of heart attacks and strokes.

In the UK cannabis is illegal and cannot be prescribed. However, it is used illegally for recreation by some people.

Testing new drugs

New drugs have to be rigorously tested before being licensed. They are tested on laboratory animals and human tissue to see if they are toxic. Then they are trialled on human volunteers.

Did you know...?

Some animals self-medicate. They eat certain leaves that they do not normally eat, to treat parasitic infections.

Key words

drug, addiction, withdrawal symptoms

▲ Cannabis products: seeds, a leaf, dried parts, and marijuana

C What are the medical benefits of cannabis?

D Why is cannabis not available on prescription in the UK?

Questions

1 Name three drugs that can be obtained from opium poppies.

2 Explain the following terms: drug; addiction; withdrawal.

3 Explain why new drugs have to be tested before they are licensed for use as medicines.

E ↓
C ↓
A* ↓

Exam tip AQA

✓ Do not fall into the trap of saying that because something is made from natural substances it is bound to be good for you. Many strong poisons come from plants.

Learning objectives

After studying this topic, you should be able to:

✔ understand why new drugs have to be tested

✔ describe how a double blind trial is carried out

Key words

Thalidomide, clinical trial, placebo, double blind trial

Did you know...?

Sometimes clinical trials do not have a placebo. If the patients are very ill it would not be ethical to give them a placebo, because they need some treatment. In this case the new drug is tested against the best current treatment. Once again, the trial is double blinded.

A Why was Thalidomide given to pregnant women?

B What is a side-effect?

Thalidomide

Between 1957 and 1961 a drug called **Thalidomide** was developed as a sleeping pill. It was prescribed to pregnant women, as it also prevented morning sickness. However, it had not been properly tested on animals, or in humans in **clinical trials**. Unfortunately it had side-effects. These are effects of the drug on the body other than the beneficial effects it is designed for. Many side-effects are minor, but thalidomide caused birth defects. The babies of women who took the drug in pregnancy had very short limbs.

Later research found that the drug interfered with genes. It prevented the normal development of limbs in the fetus.

The drug was then banned. However, since the 1980s it has been used in some countries in the treatment of leprosy. Unfortunately doctors have not always checked that patients are not pregnant. As a result, in those countries, children have more recently been born with very short limbs.

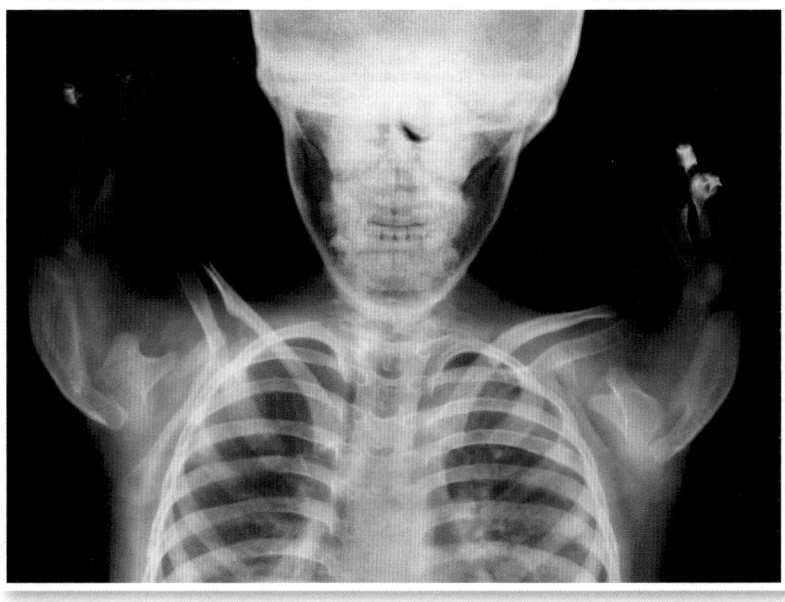

▲ X-ray of upper chest and arms of a baby. The baby's mother was given Thalidomide when she was pregnant. This caused the baby to have very short arms because the arm bones did not develop properly.

Because of this tragedy, new drugs have to be rigorously tested in clinical trials before being licensed.

How new drugs are tested

- New drugs are tested in laboratories, on human tissue and animals, to see if they work and to find out how toxic they are.
- If they pass these tests the drugs are tested on humans in clinical trials.
- At first very low doses of the drug are given to volunteers.
- Then doses are increased to find the dose that works best.
- Volunteers are divided into two groups. The control group is given a **placebo** (dummy pill) and the experimental group is given the real drug.
- Neither the doctors nor the patients know who is getting the placebo and who gets the real thing, until the end of the trial. This is called a **double blind trial**. It makes the trial fair.
- At the end of the trial the two groups are compared to see if there is any real difference between them.
- If the drug makes a real difference and causes no harm it is licensed for use.
- However, if some new side-effects occur, some drugs are recalled even when they have been used on many people.

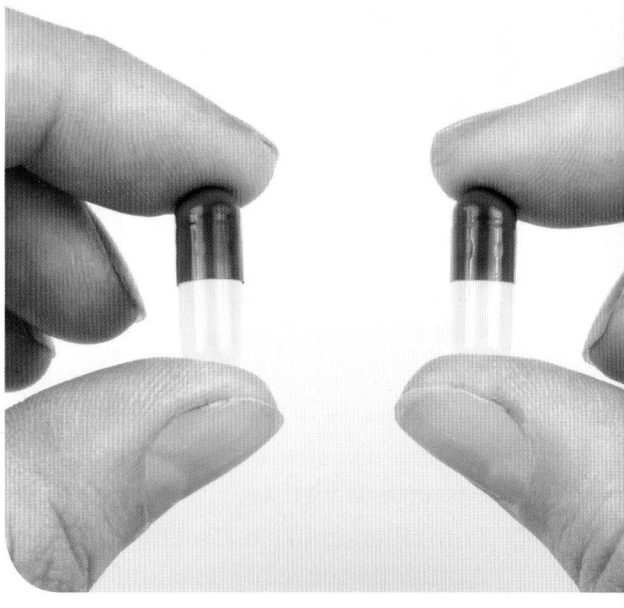

▲ One of these capsules contains real medicine. The other is a placebo. It looks like the real thing but does not contain medicine. Sometimes people feel better when taking a placebo, because they believe they are being treated.

▲ This boy is taking a tablet of omega-3 oil. He is part of a clinical trial to see if omega-3 can improve children's brain function. One group took the real tablets twice a day for eight weeks. The control group took a placebo twice a day for eight weeks. Members of each group had memory tests. Their reaction times and attention spans were also measured.

Questions

1. What were the effects of Thalidomide when given to pregnant women?

2. Explain why new drugs have to be tested before they are licensed for use as medicines.

3. How are new drugs tested?

4. Discuss why in a double blind trial (a) the patient and (b) the doctor does not know whether they are being treated with the real drug or a placebo.

Course catch-up

Revision checklist

- Your body type and activity levels decide how much energy you use.

- To stay healthy, you need a balanced diet. This provides enough of each nutrient and the right amount of energy.

- Pathogens cause disease by multiplying inside us. They can be bacteria or viruses.

- Painkillers make us feel better, but do not kill pathogens. Antibiotics cure bacterial infections until mutation and natural selection let resistant strains develop.

- Our bodies keep most pathogens out. White blood cells fight infections by ingesting pathogens, producing antibodies or making antitoxins. Vaccines provide immunity by preparing white blood cells to produce antibodies faster.

- Our brains process signals from sensory cells and decide how to respond. We also have fast, automatic reflex actions which protect us from danger.

- We keep our temperature, water content, ion content, and blood glucose steady. Chemical messengers called hormones help by coordinating the actions of different cells.

- Hormones regulate a woman's menstrual cycle and are used as contraceptives.

- Hormones also make IVF possible. Controlling fertility brings benefits and problems.

- Plants also use hormones to respond to change.

- Drugs alter the way your body or brain works. Some people use recreational drugs to alter their mood, but drugs can be harmful and/or addictive.

- New medicinal drugs need testing to show they work and aren't harmful. Double blind trials are used, which compare drugs with placebos. Doctors and patients aren't told who took which, so the results cannot be biased.

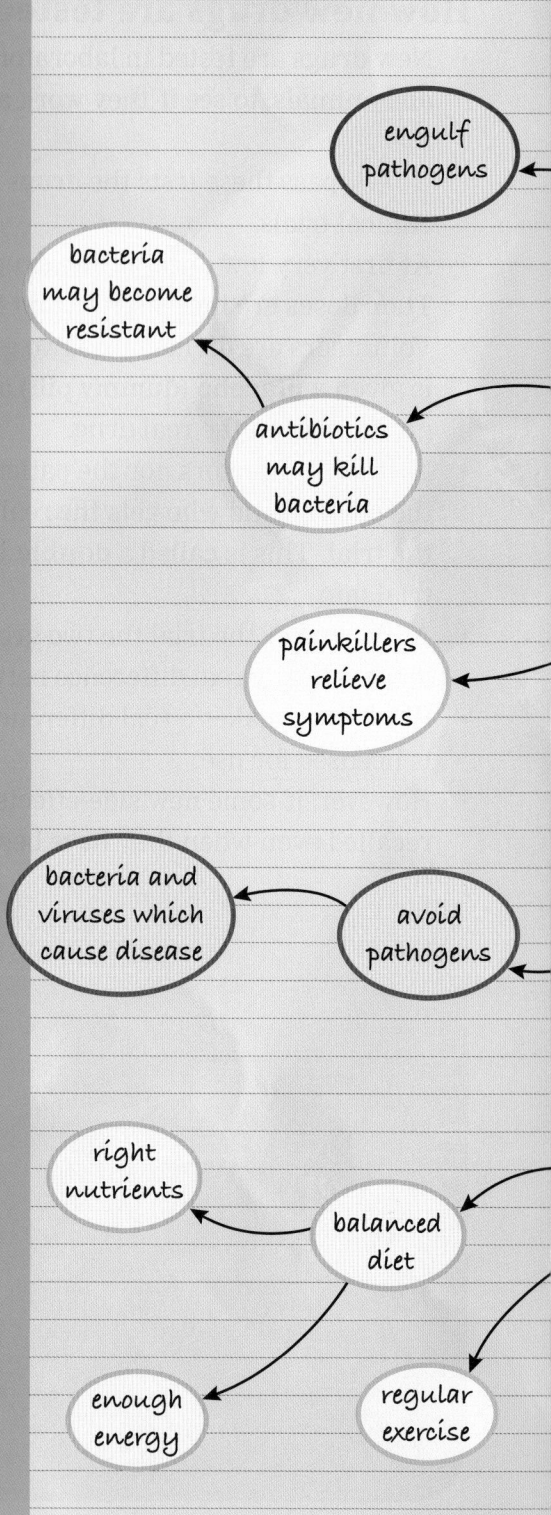

engulf pathogens

bacteria may become resistant

antibiotics may kill bacteria

painkillers relieve symptoms

bacteria and viruses which cause disease

avoid pathogens

right nutrients

balanced diet

enough energy

regular exercise

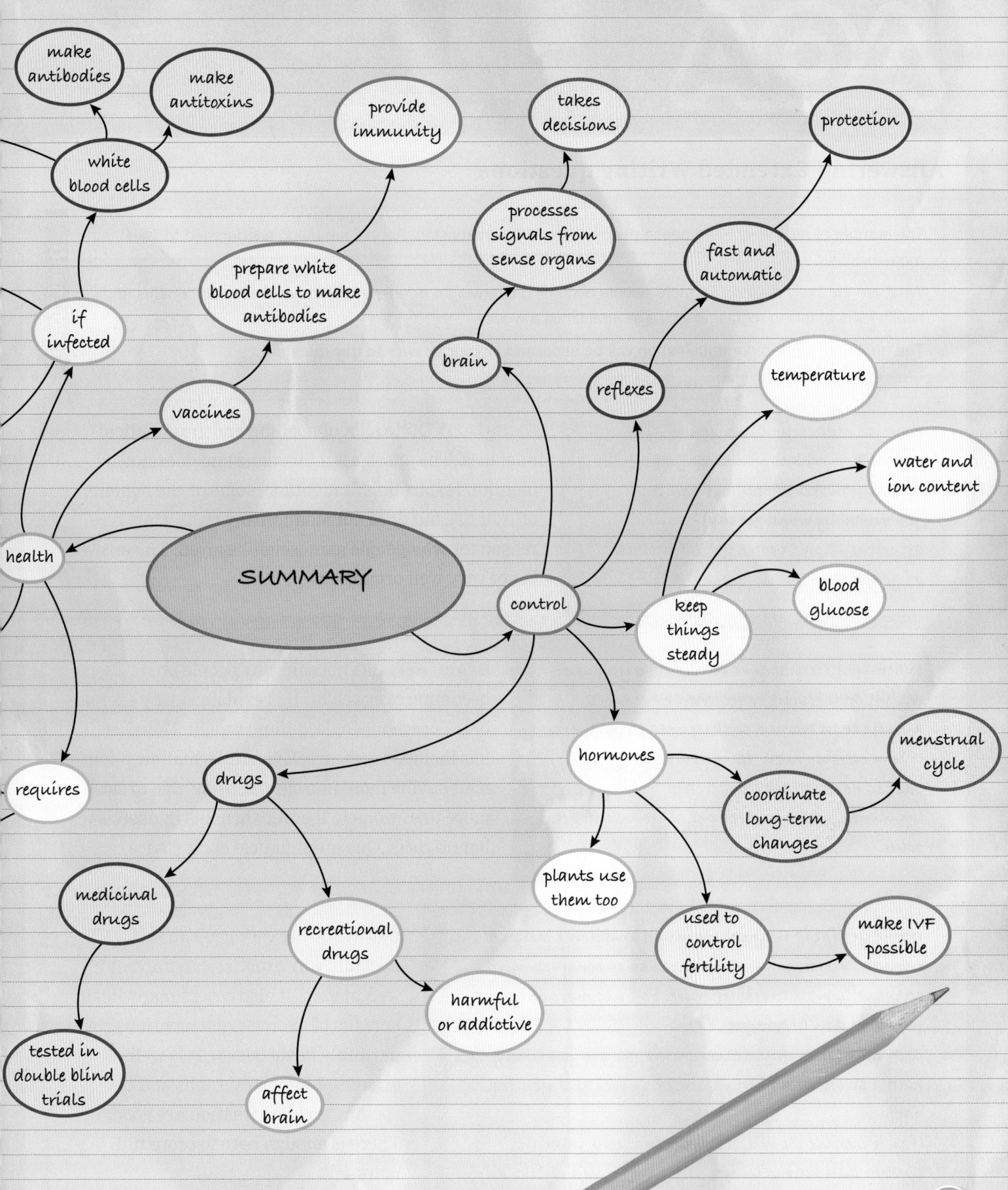

make antibodies

make antitoxins

white blood cells

provide immunity

takes decisions

protection

prepare white blood cells to make antibodies

processes signals from sense organs

fast and automatic

if infected

brain

reflexes

temperature

vaccines

water and ion content

health

SUMMARY

control

keep things steady

blood glucose

requires

drugs

hormones

menstrual cycle

coordinate long-term changes

medicinal drugs

recreational drugs

plants use them too

used to control fertility

make IVF possible

tested in double blind trials

harmful or addictive

affect brain

Answering Extended Writing questions

QUESTION

The number of overweight and obese people in the UK is increasing. People are being encouraged to take more exercise.

What are the likely reasons for more people becoming overweight or obese? What are the health benefits and health risks of taking regular exercise several times a week?

The quality of written communication will be assessed in your answer to this question.

Exercise makes you use your mussels. It makes you strong. It makes you feel happy and healthy. You probably wont get diabetic and arthritus. Some people get fat because of genetics.

G–E

Examiner: Quite a lot of spelling and grammatical mistakes. The answer is a bit vague and does not mention respiration. It mentions some of the benefits of exercise but does not mention any risks. Only one reason for why people gain weight is given. The answer is not very well organised.

When you exercise you burn fat so you lose weight. You respire more. Men have more muscle than women so they need to eat more. If you exercise most days you wont get cancer, heart attacks or strokes. Your immune system is stronger so you wont get colds.

D–C

Examiner: Has not said why people are becoming overweight. Has explained why exercise helps prevent people gaining weight. However, it has included some irrelevant information, about men having more muscle and being able to eat more than women. It covers benefits of exercise but no risks. One grammatical error (twice).

People get fat if they eat too much and don't exercise enough. Exercise makes your muscles contract more so the cells need to respire more. You use more of your food and don't store fat.

Exercise makes muscles stronger. You are less likely to have a stroke or a heart attack or cancer and you feel happier.

If you exercise too much it weakens your immune system and harms joints.

B–A*

Examiner: A very good answer. It explains why exercise can prevent weight gain. It also covers many health benefits of exercise and mentions some harmful effects (risks). It is well organised into paragraphs and the spelling and punctuation are good. There is enough here to score full marks.

Exam-style questions

1 The diagrams show two organisms that can cause disease.

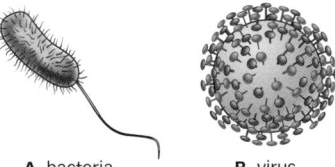

A bacteria   **B** virus

Answer the following questions by writing A, B, neither, or both.

A01 **a** Which could be called a pathogen?

A01 **b** Which is destroyed by antibiotics?

A01 **c** Which is destroyed by painkillers?

A01 **d** Which would be able to multiply rapidly inside you?

A01 **e** Which multiplies inside cells?

2 Cannabis changes how people see and hear things, and makes them feel relaxed. But could it be doing long-term harm? Scientists used an MRI machine to compare the brains of users and non-users. They were all the same age. The 15 users had smoked cannabis five times a day for ten years. The brain area that deals with emotions was 12% smaller in users' brains.

A02 **a** Why did the scientists choose long-term cannabis users for their study?

A02 **b** Why was it important to compare users and non-users of the same age?

A03 **c** Can we be sure that the cannabis has changed their brains?

A03 **d** Suggest a way to get stronger evidence that cannabis shrinks parts of the brain.

A03 **e** Suggest one ethical problem you might have if you carried out this experiment.

G–E   *D–C*

3 Many women use hormones to control their fertility.

A01 **a** Explain what a hormone is.

A01 **b** Why do hormones circulate in your blood?

A01 **c** Explain why oestrogen and progesterone are used in oral contraceptives.

A01 **d** Give two advantages and two disadvantages of using oral contraceptives.

A02 **e** Suggest why the composition of the pills has changed since they were first used.

B–A*

Extended Writing

4 Dev cut his hand in the garden. His
A02 finger is red and swollen. It's infected. Explain how Dev's white blood cells will try to combat the microbes.

5 Kevin is taking part in a drug trial. It is
A02 a double blind trial of a new acne cream. Twenty people are taking part and will stay at the test centre for three weeks. They all have acne. Write a set of instructions for the nurses running the trial.

6 Mary and John want to have children
A02 using IVF. Explain how and why IVF
A03 is used, and state arguments against its use.

G–E   *D–C*   *B–A**

A01 Recall the science

A02 Apply your knowledge

A03 Evaluate and analyse the evidence

B1 Part 2

Surviving and changing in the environment

Why study this unit?

You have seen how organisms are adapted to their environment. In this unit you will continue to look at how organisms survive. Organisms gradually change – better characteristics making them more suited to their environment.

In this unit you will look at how the distribution of organisms will change, as natural conditions change. These changes in conditions may be the result of human pollution. The movement of energy through the living world, and the cycling of elements between the living and, non-living world, will also be looked at.

You will also look at the process of reproduction in animals, and the ways in which humans can manipulate reproduction. This will lead on to the controversial issues surrounding cloning. Finally, the ideas of groups within the variety of life, and an explanation of how nature selects characteristics most suited to their environment via natural selection, will be investigated.

You should remember

1 Organisms depend on each other for their survival.

2 Feeding relationships are shown in food chains and webs.

3 Energy is passed through the food chain by animals feeding.

4 Plants and animals are sorted into groups based on their characteristics. This is called classification.

5 There is variation between individuals of the same species.

6 Plants and animals reproduce to produce new individuals.

One of the most majestic bears on Earth is the polar bear. The polar bear and its closest relative, the brown bear, diverged about 150 000 years ago. Males can be 2.4 metres long and can weigh up to 680 kg, making it the largest land carnivore alive today. Our increasing demand for energy has led to global warming which has caused the ice the polar bear lives on to melt. Their numbers are now in decline because of this change in habitat, and also due to hunting. In the past the polar bear has adapted to changes in climate, as natural climate change occurs slowly. But biologists are worried that the bear will not have time to adapt to the change this time.

Learning objectives

After studying this topic, you should be able to:

- ✔ know that adaptations of organisms help them survive
- ✔ identify and explain adaptations of animals

Did you know...?

Some Amazonian tribes use the poison from the skin of poison dart frogs to tip their arrows. This is one of the most deadly poisons known. They use the arrows to kill monkeys.

What are different environments like?

Conditions in different environments can vary greatly. Deserts are hot and dry environments; the Arctic is cold. There are animals that can survive in each of these environments, but they must be adapted to survive. An **adaptation** is a feature of an animal's body which helps it to live in its environment.

Adapted to compete

Animals have to be adapted not only to the physical environment, but also to cope with other organisms in their environment. Some animals are **camouflaged** to blend in with their environment, so predators or prey do not notice them. Other animals are poisonous and have bright warning colours to prevent them being eaten. An example is poison dart frogs. The giraffe's tongue is adapted to stretch out and pull leaves from between the thorns of the acacia tree.

Adaptations to cold environments

Adaptation	How this aids survival
small ears	the surface area of the ear is reduced, and so less heat is transferred to the environment
thick white fur	insulates the body against the cold, and camouflages the bear
sharp teeth	to kill prey
strong legs long legs	contain large muscles which contract so the bear can run on land or swim in water
big feet with fur on the soles	spread the load of the animal on the snow or ice; fur helps grip and helps insulate against the snow
claws	for killing and holding prey
blubber below the skin	a thick layer of fat which insulates against heat transfer to the environment; the stored fat can also be used for respiration to generate heat
large body size	reduces the relative surface area, and so reduces heat loss
fins	balance the whale during swimming
muscular tail	contains large muscles which contract to generate movement during swimming

A State three adaptations of the polar bear that only help it survive if conditions are cold.

B Penguins live in the Antarctic. Describe five features of a penguin that help it to survive there.

Adaptations to hot environments

Adaptation	How this aids survival
hump of fat	fat is stored in one place, which reduces all-round insulation; fat can be broken down to release water
bushy eyelashes	stop sand entering the eyes
nostrils which close	prevent breathing in of sand
body tolerance to temperature changes	does not need to sweat so much when hot
long legs	lift body off hot sand
large feet	spread the load, stop the camel sinking into the sand
thin fur	less warm air is trapped, reducing insulation
large ears	lose heat by radiation, and are used to fan the body
large body size	can knock over plants and shrubs for food
wrinkled skin	increases the surface area from which to lose heat
trunk	allows the elephant to suck up water to drink, and to spray water over the body to cool itself
large feet	spread the load and stop the elephant sinking into the mud

C   Sort the adaptations of the elephant and camel into two lists – those that help them survive hot conditions, and those that are not related to the heat.

D   Explain how the difference in ear size between elephants and polar bears helps the two animals survive in their different environments.

Key words

adaptation, camouflage

Questions

1   A close relative of the elephant is the woolly mammoth. It lived in cold environments. Suggest two features that differ from those of the elephant and helped the mammoth survive.

2   Explain how the features you identified in Question 1 would aid the mammoth's survival.

3   The distribution of camels is limited to the desert. Explain why the camel is well adapted to life in the desert.

4   Explain why slow-moving camels are not well adapted to live in the community of animals on the African savannah.

E
C
A*

Exam tip

✓   When looking for adaptations, notice particular features of an animal, and try to suggest how the animal uses these features to survive.

Learning objectives

After studying this topic, you should be able to:

✔ know that plants show adaptations to hot and cold environments

✔ understand that the adaptations of a plant determine where it can grow

Key words

pore, needles, surface area, spines

A What is the advantage to the oak tree of its large green leaves?

B Name the advantage to the pine tree of reducing its leaves to needles.

C Suggest a disadvantage to the pine tree of reduced leaves.

Exam tip **AQA**

✔ Think about how a plant is adapted to its environment, and how its adaptations restrict its distribution.

Plants are adapted too

Plants are found in all sorts of environments. They need to be adapted just as well as animals to survive. Plants show a range of adaptations, most of them to do with absorbing light and retaining water. Leaves absorb light, and water is lost from the surface of the leaves through **pores**.

Adaptations to hot and cold environments

Northern pine forests	Temperate forests
Pine tree	Oak tree
Conditions	**Conditions**
For much of the winter the water freezes in the soil.	Cold and not much light in winter; sunny and moist in summer.
Adaptations	**Adaptations**
Leaves are reduced to **needles** to reduce the **surface area**; this reduces water loss.	Leaves fall from the trees in autumn; there is not much light for photosynthesis, so leaves have no function in the winter.
	Leaves have a larger surface area to make the most of the summer sun.
Thick wax on the surface of the leaf also reduces water loss.	Wax is thinner on the leaf surface, as there is plenty of water in the soil for most of the year.

Adaptations to dry environments

Beaches and sand dunes	Deserts
Marram grass	Cactus
Conditions	**Conditions**
Rainwater quickly drains through sand. Can be hot and windy.	Very little water in the sandy soil; very low rainfall. Air is hot and dry.
Adaptations	**Adaptations**
Leaves are long thin spikes, which reduces their surface area; this reduces water loss. Leaves are rolled; leaves lose water through pores, and rolling keeps these pores on the inside of the roll. Waxy layer on the outside of the leaf to reduce water loss. Deep root system to absorb water, and to anchor the plant against the wind.	Leaves are **spines**, which reduces their surface area; this reduces water loss. Spines are less likely to be eaten by animals. Wax on stem reduces water loss. Shallow root system to cover great areas, and to absorb water when it does rain. Stems are swollen to store water. Many stems have grooves which can expand and flatten out when the stem fills with water.

The acacia tree is common on the African savannah. These trees are heavily grazed by animals such as giraffes. In order to reduce the damage caused by grazing, acacias have a number of adaptations.

- They have developed large thorns which discourage the giraffes.
- The thorns provide a home to colonies of stinging ants.
- The leaves release a nasty poisonous chemical which giraffes dislike.

▲ A giraffe browsing an acacia tree

D The pine tree and the cactus both show similar adaptations of their leaves. Explain why the same adaptation helps plants survive in two different environments.

E Beaches are often windy environments. Explain how the adaptations of marram grass help it survive the windy conditions.

Questions

1 State the adaptations that help marram grass live in dry soil.

↓ E

2 Explain why the dandelion isn't well adapted for living in the desert.

↓ C

3 Suggest a reason why flowering plants tend to reproduce more in the warmer summer months.

4 The acacia tree has a very long tap root. Explain how this adaptation helps it to survive on the African savannah.

↓ A*

Learning objectives

After studying this topic, you should be able to:

✔ know that many extreme conditions are found on Earth

✔ know that organisms are found in most environments, and show adaptations to those extreme environments

Key words

extremophiles

▲ A red mangrove tree growing in an estuary in the Caribbean

A **What is an extremophile?**

B **Explain why plants which live on salt marshes in estuaries need to be adapted to deal with salty conditions.**

Living on the edge

There are many places on Earth with very difficult conditions – too extreme for humans to live there. But some organisms do live in these places. They are called **extremophiles**. These organisms are often fascinating because they have interesting and unusual adaptations.

Coping with salt

Mangrove trees are common plants at the water's edge on tropical coasts. Their roots are permanently in sea water, with salt levels that would kill most land plants. So how do they survive? Water is lost from their leaves and, as in most plants, this causes the plant to draw in large amounts of water through the roots. But mangroves have special roots that do not allow salt to enter as the water is absorbed.

Under pressure

Many marine animals, such as whales, are able to dive to astonishing depths. Deep in the ocean, the pressure increases. It would crush a human, but whales can survive because their bodies are highly adapted to cope. Their lungs are smaller than expected for an animal so big, and they do not fully inflate them before diving. This makes them less buoyant. Their muscles can store large amounts of oxygen, which allows them to remain underwater for long periods.

▲ A sperm whale is adapted to life deep in the ocean

Microbes far and wide

Microorganisms have such a wide range of adaptations that there are microbes of one kind or another living in almost all the conditions found on Earth. Bacteria are found in the depths of the ocean and on the peaks of mountains. They span cold environments such as Arctic wastes to the hot springs of Yellowstone Park in North America.

Thermophiles

Thermophiles are bacteria that are adapted to living in very high temperatures. They have been found living in environments above 80 °C, such as in volcanic areas, hot springs, and geysers. They often have bright colours.

Thermophilic bacteria are adapted to survive in these hot conditions by having special enzymes. Most enzymes are destroyed at high temperatures, but these particular enzymes are able to withstand the hot temperatures. These bacteria are useful in industry. The heat-resistant enzymes are used in genetic engineering.

Bacteria at the bottom of the sea

Biologists have discovered bacteria living under great pressure at the bottom of the ocean. These bacteria are adapted by having an equally high pressure inside their cells, to balance the pressure of the water around them. They are biologically important because they can carry out reactions which release energy for them to survive.

When scientists look for life on other planets, they tend to look for bacteria, because these are the organisms that are most likely to survive the extreme conditions.

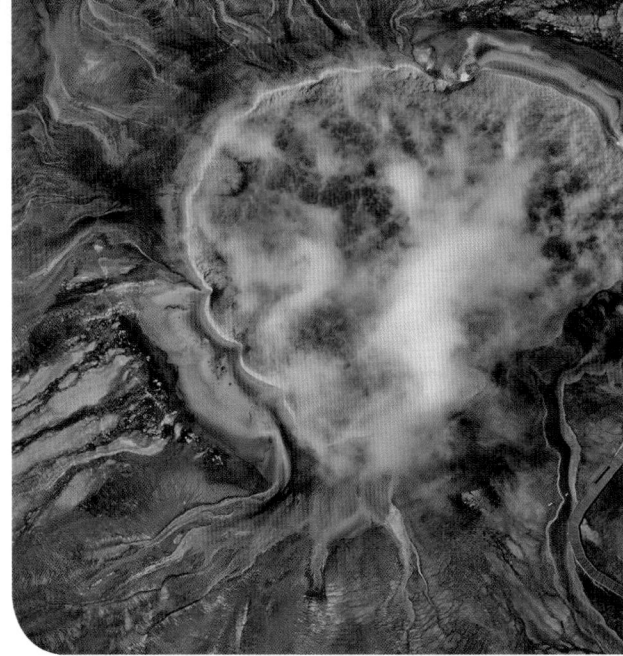

▲ This lake in Yellowstone Park contains many thermophilic bacteria which colour the water

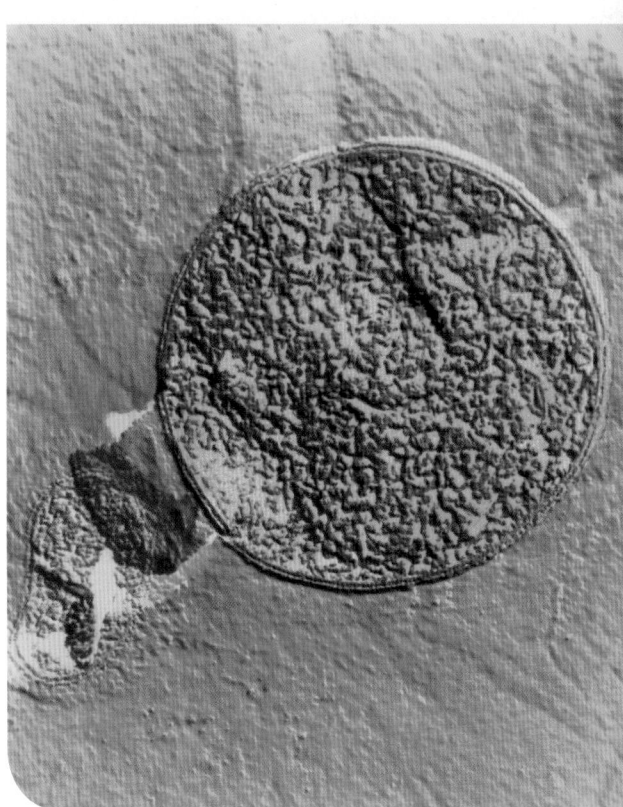

▲ Thermophilic bacteria are often brightly coloured

Questions

1 What is a thermophile?

2 Describe what adaptations are shown by bacteria that live in hot-water geysers.

3 Explain why thermophilic bacteria are of use to scientists who carry out cell reactions at high temperatures.

E
↓
C
↓
A*

Learning objectives

After studying this topic, you should be able to:

✔ know that organisms compete with each other for resources, and that this can affect their distribution

Key words

resource, competition, population

Did you know...?

Humans are probably the greatest competitors of all time. This has led to many other species losing out. Dodos and Tasmanian tigers have both become extinct because of humans.

A Why do farmers and gardeners remove weeds from around their plants?

B Explain why bluebells only grow in the spring in a British woodland.

What do plants and animals compete for?

There is a limited supply of **resources** for plants and animals. Nearly all living things are locked in a battle to get enough materials and other resources to survive. This fight for resources is called **competition**. The availability of resources can affect the distribution of an organism – that is, where it lives.

Plants compete for:
- light
- space
- water
- minerals
- carbon dioxide.

Animals compete for:
- food
- space or territory
- mates
- water.

A population is the number of organisms of a particular species in a named area. For example, the woodlice living under a stone are a **population**. If conditions are good and the woodlice reproduce, the population gets larger. If conditions are bad and some woodlice die out, the population gets smaller. Competition affects the size of populations.

Competition between plants

Carbon dioxide: plants need this for photosynthesis. The air contains only 0.03% carbon dioxide. The massive canopy of tree leaves absorbs carbon dioxide, so there is less available under the tree for ground plants.

Light: the energy supply for photosynthesis. The tree leaves absorb some light, and not much light passes through, so it is too shady under a tree for many plants to grow.

Space: the tree roots take up most of the space in the ground, leaving little room for ground plants.

Water: used in photosynthesis and to cool the plant. The large tree will absorb most of the water in the soil, leaving little for the smaller ground plants.

Soil minerals: needed to keep the plant healthy. These are absorbed in water through the roots. Again, the tree can absorb far more minerals than the small plants.

▲ Plants do not move around, so they can only live in places where resources are available. If they cannot compete with other plants for these resources, they will not be able to survive there.

Competition between animals

Animals usually compete with each other for food. For example, some coastal birds compete for the same food supply, and they must find a solution for the different species to survive together.

> **C** Name the most common resource for which animals compete.

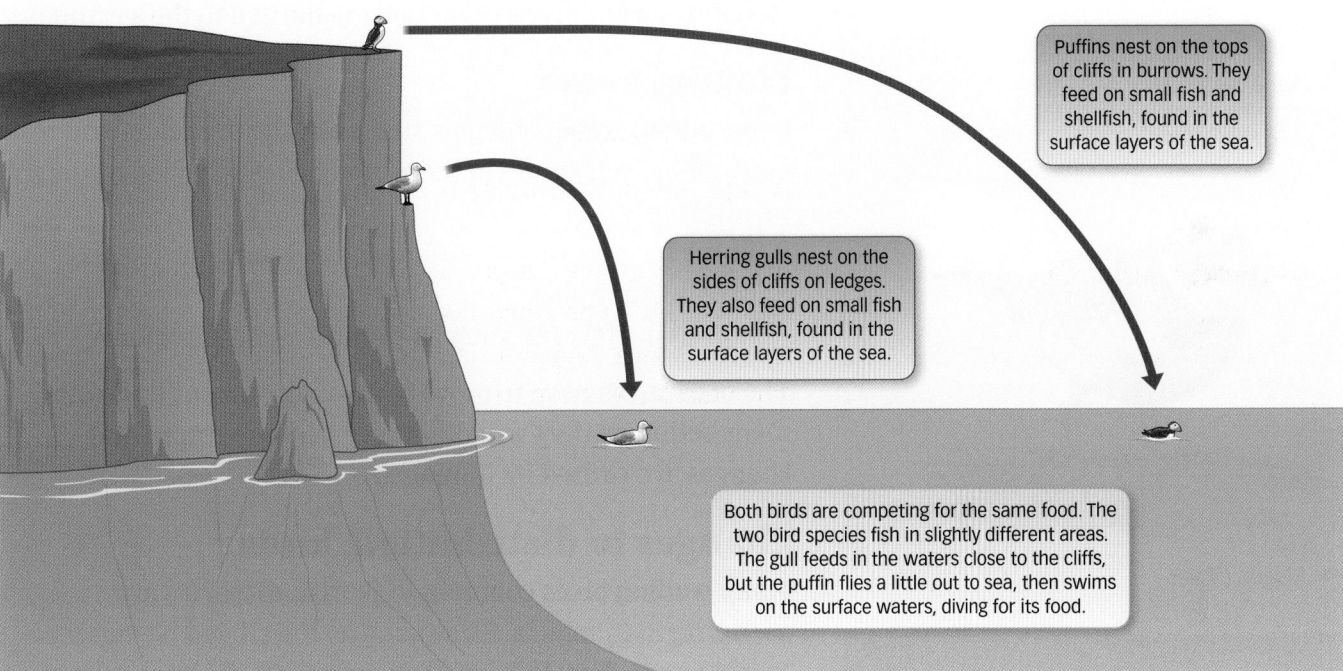

Puffins nest on the tops of cliffs in burrows. They feed on small fish and shellfish, found in the surface layers of the sea.

Herring gulls nest on the sides of cliffs on ledges. They also feed on small fish and shellfish, found in the surface layers of the sea.

Both birds are competing for the same food. The two bird species fish in slightly different areas. The gull feeds in the waters close to the cliffs, but the puffin flies a little out to sea, then swims on the surface waters, diving for its food.

▲ Competition between the puffin and the herring gull

Questions

1 Explain why plants need light to survive, but animals do not.

2 Explain what will happen to the ground plants in a woodland when a large tree dies and falls.

3 Red squirrels are a native species in the UK. Around 1900 a close relative, the grey squirrel, was introduced. The grey squirrel is a better competitor. Describe the effect the grey squirrel's introduction had on the red squirrel population.

↓E ↓C ↓A*

> **D** If humans fished the coastal surface waters, explain what would happen to the population of puffins.

Exam tip AQA

- ✔ When two populations compete with each other, the number of individuals in each population is reduced.

Learning objectives

After studying this topic, you should be able to:

✔ know that the distribution of organisms is affected by the environment

✔ discuss the change in distribution of named species

Key words

distribution

▲ Ringed plover

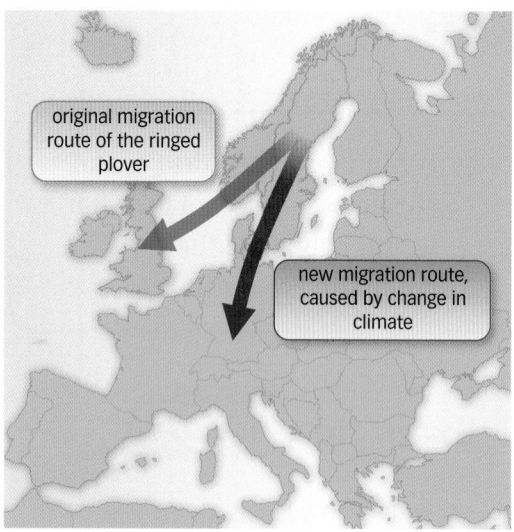

original migration route of the ringed plover

new migration route, caused by change in climate

▲ Climate change has resulted in lower numbers of ringed plover in the UK

Fit for the job

A well-adapted animal survives well in its environment. It is fit for the job.

Elephants are well suited to the grassy African savannah, so there are large numbers of elephants on the savannah. They cannot necessarily survive well elsewhere. Elephants are not adapted to survive further north in Africa, where there are deserts, because they cannot survive long periods without water, and there would not be sufficient food for them. So the distribution of African elephants is limited to the savannahs.

Moving home

If the environment changes suddenly, the organisms there may not be well suited to these changes. Such changes might result from:

- living factors, such as humans, predators, or disease-causing microbes
- non-living factors, such as rainfall and temperature.

The organisms have to move to an environment that suits them better, or they will die. Many fast environmental changes are caused by humans.

Changes in distribution: birds

Many wading birds common to British estuaries are declining in numbers. Birds such as the ringed plover used to migrate to Britain from the north for the milder winters. However, as the climate has become warmer throughout Europe during the winter, they have been moving to mainland Europe where the conditions are better. The result is much lower numbers of ringed plover in Britain.

Biologists working for the Royal Society for the Protection of Birds (RSPB) have been concerned about this trend. They have used computers to predict bird **distributions** in the future. The results have suggested that the average bird species distribution will shift nearly 550 km north-east by the end of this century. This is equivalent to the distance from Plymouth to Newcastle. The distribution is also likely to be reduced. This may result in some birds declining in numbers or even dying out altogether.

Human developments, such as the flooding of bays like Cardiff Bay, result in the loss of natural homes for wading birds. During such developments, the planners make provision for the creation of nature reserves close to the development, where bird species can be relocated.

▲ An artificial wetland habitat created as a nature reserve at Cardiff Bay, Wales

> A What factors restrict the distribution of an organism to a particular area?
>
> B Suggest a reason why beech trees are restricted to growing mainly in the southern half of the UK.
>
> C What might be the impact of global warming on the distribution of beech trees?

Changes in distribution: bees

Bees are another species that have been affected by the changing environment. They are declining in number. Biologists believe that there are four reasons for the decline:
- a virus that attacks the bee larvae
- summers are cooler and damper
- the use of agricultural chemicals
- an increase in air pollution.

Questions

1 Describe what might cause the distribution of an animal species to change. ↓ E

2 State what might happen to a population of birds if they did not move in response to changes in the environment.

3 The ice caps are melting due to global warming. Explain what this means for the distribution of animals such as the polar bear. ↓ C

4 Why is it important that developers create nature reserves for species when developing an area? ↓ A*

▲ The honey-bee is declining in the UK

Learning objectives

After studying this topic, you should be able to:

✔ know that organisms are affected by pollution

✔ know that organisms can be used to indicate levels of pollution

Key words

pollutant, indicator species

▲ A biologist collecting water samples for testing

Pollution and biodiversity

Pollutants are harmful substances that humans add to the environment. Pollutants have an impact on the number and type of organisms that can survive. Generally, in more polluted areas, fewer species can survive. In some cases particular species will be the main or only survivors, and may even be adapted to cope quite well with the pollution. The presence of these species show biologists that the area is polluted. They are called **indicator species**.

Measuring pollution levels

Biologists use the following two methods to measure the levels of pollution:

1. Non-living indicators – biologists measure a variable such as the temperature, pH, or oxygen level in the environment.
2. Living indicator species – the presence of species adapted to survive in polluted conditions shows that the area is polluted.

Biological oxygen demand

Water in streams and lakes contains oxygen, which allows fish and other organisms to survive. Pollution can lead to lower oxygen levels. These are generally caused by bacteria, which use up the oxygen. The amount of oxygen being used by the bacteria is called the biological oxygen demand (BOD). Biologists can measure the levels of oxygen in water with an oxygen meter. In clean water the BOD is low, but in polluted waters the BOD is higher. The BOD is a measure of pollution.

Other common variables that biologists use to measure pollution in different habitats include

- temperature, measured with a thermometer
- rainfall levels, measured by collecting rainwater and recording the volume.

An indicator species in water

Rat-tailed maggots are invertebrate animals that are adapted to survive in water with very little oxygen (a high BOD). These maggots have a long, tail-like tube which is hollow. It acts like a snorkel, allowing the maggot to take in air containing oxygen from above the polluted water. This is why the maggot thrives in these conditions, and is an indicator species of polluted water.

Lichens: another indicator species

Burning fossil fuels releases many chemicals into the air, including sulfur dioxide. This causes air pollution, which reduces the variety of organisms that can survive in the area. Lichens are one group of living things that act as indicator species of air pollution.

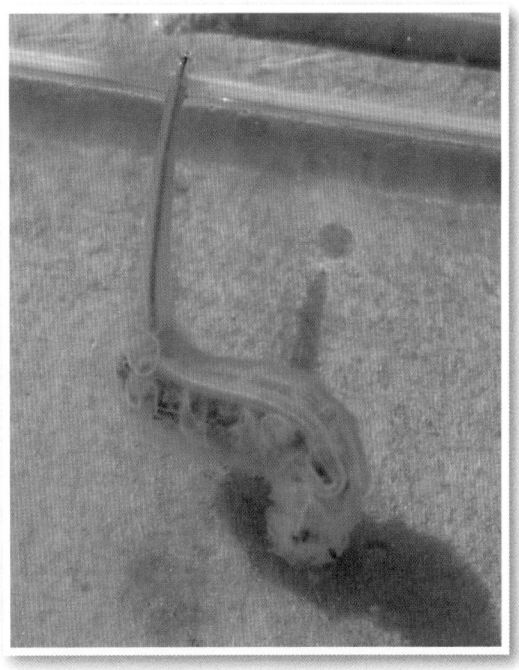

▲ Rat-tailed maggot

> A Explain the importance of indicator species.
>
> B The National Rivers Authority (NRA) is responsible for keeping fishing rivers clean. Suggest what they might look for in water samples from rivers.

Some lichens can cope with high levels of pollution, and are found in cities. Other lichens cannot grow there, and are only found in areas with clean air away from cities and motorways. Lichens are great indicators of the level of air pollution in an area.

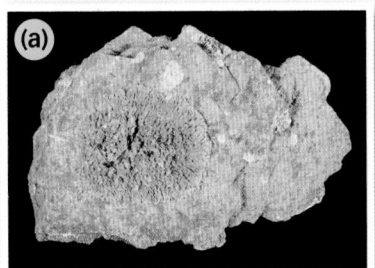
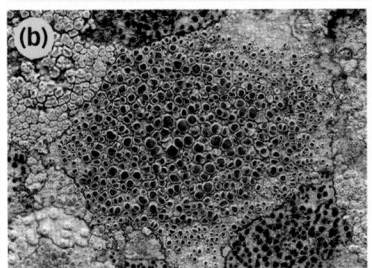

▲ These lichens are indicators of
(a) polluted air
(b) moderate pollution
(c) clean unpolluted air

Questions

1 If a water sample was tested and showed a high BOD, what would this tell you about the water? E

2 Explain how you could measure the average rainfall in your school playground per day durng December.

3 Rat-tailed maggots are poor competitors. Suggest why they do not survive well in clean water.  C

4 Describe an experiment you could carry out to show how lichens can be used to indicate the levels of pollution as you move out of a city.  A*

Learning objectives

After studying this topic, you should be able to:

✔ understand how food chains and pyramids of biomass show the feeding relationships between organisms

✔ appreciate the ways in which scientists work

▲ Pyramid of numbers for the food chain on the African savannah

▲ Pyramid of biomass for the food chain on the African savannah

Food chains

A **food chain** shows how the organisms living in an environment are linked by which organisms eat which others. The food chain shows the flow of food and energy from one organism to the next. Each organism is linked by an arrow which shows the direction of flow of food and energy. Here is a food chain for organisms living on the African savannah.

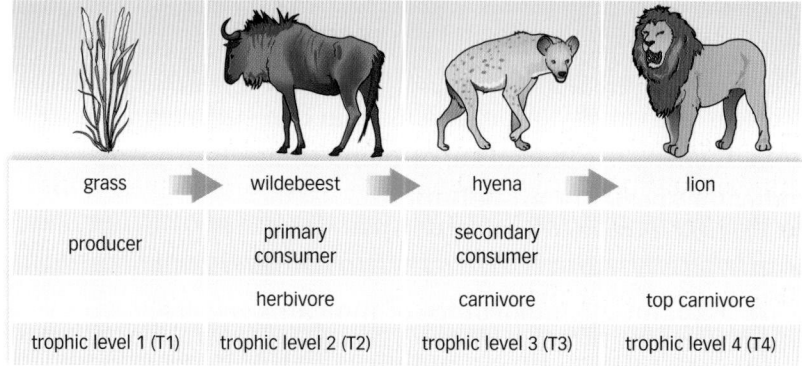

grass	wildebeest	hyena	lion
producer	primary consumer	secondary consumer	
	herbivore	carnivore	top carnivore
trophic level 1 (T1)	trophic level 2 (T2)	trophic level 3 (T3)	trophic level 4 (T4)

Biologists often study how many organisms there are in each link of a food chain. For example, there are large numbers of small grass plants at the start of the chain shown on the right, but only a few lions at the end of the food chain. These numbers can be plotted in a pyramid of numbers. You can see in the pyramid on the left that the number of organisms decreases at each link in the food chain.

Pyramids of biomass

In a **pyramid of biomass**, biologists plot the biomass of the organisms at each link of the food chain rather than the number of organisms. To calculate the biomass, they multiply the number of organisms at each link of the food chain by the dry mass of one organism. The biomass for each link in the chain can now be plotted as a pyramid of biomass.

> A Describe what happens to (a) the number and (b) the size of organisms as you pass along a food chain.
>
> B What does a pyramid of biomass show?
>
> C Sketch a likely pyramid of biomass for the following food chains:
> grass → caterpillar → blue tit → hawk
> grass → impala → cheetah → fleas

How scientists work

A case study: lions under threat

Pyramids give us a picture of the state of the environment. If the pyramids change shape over time, this might suggest a problem in the environment. Over the last 60 years, scientists have been concerned about the falling numbers of lions on the African savannah. They wanted to know the cause.

Step 1: analyse the data

Scientists have collected data over the last 60 years for lion food chains and plotted pyramids.

The scientists realised that all the organisms have fallen in number, but particularly the lions. There were 500 000 lions in 1950 but only 20 000 in 2010.

Step 2: interpret the data

The scientists identified a relationship between the falling numbers and human activity on the savannah. They suggested that the fall in numbers might be due to

- habitat destruction by humans to build towns, which reduces the numbers of all species
- hunting, which has a specific effect on the lion population.

Step 3: use the information to inform decisions

Following these findings, people around the world who care about wildlife need to develop plans to protect the lion before it becomes endangered.

Key words

food chain, pyramid of biomass

Did you know...?

The wild lion population of Kenya could become extinct by 2030 if nothing is done to stop its decline.

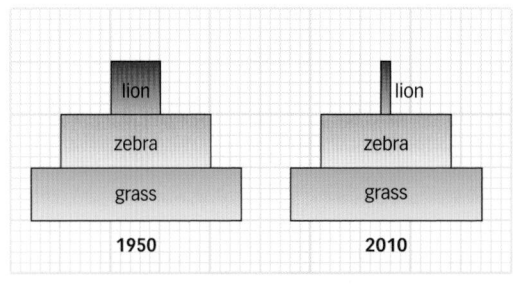

▲ Pyramids of numbers for an African savannah food chain, 1950 and 2010

Questions

1 State what happens to the biomass of organisms as you move along the food chain. ⬇ E

2 Describe how the scientific community makes use of food pyramids. ⬇ C

3 Construct a pyramid of biomass for the following food chain. Use graph paper where one small square will represent 100 kg. ⬇ A*

grass	→	zebra	→	lion
1 000 000		100		4
0.1 kg		300 kg		250 kg

Exam tip

✓ You can think of a food pyramid as a graph turned on its side.

Learning objectives

After studying this topic, you should be able to:

✔ know that energy is lost at every link in the food chain

✔ appreciate that farming methods try to reduce energy loss

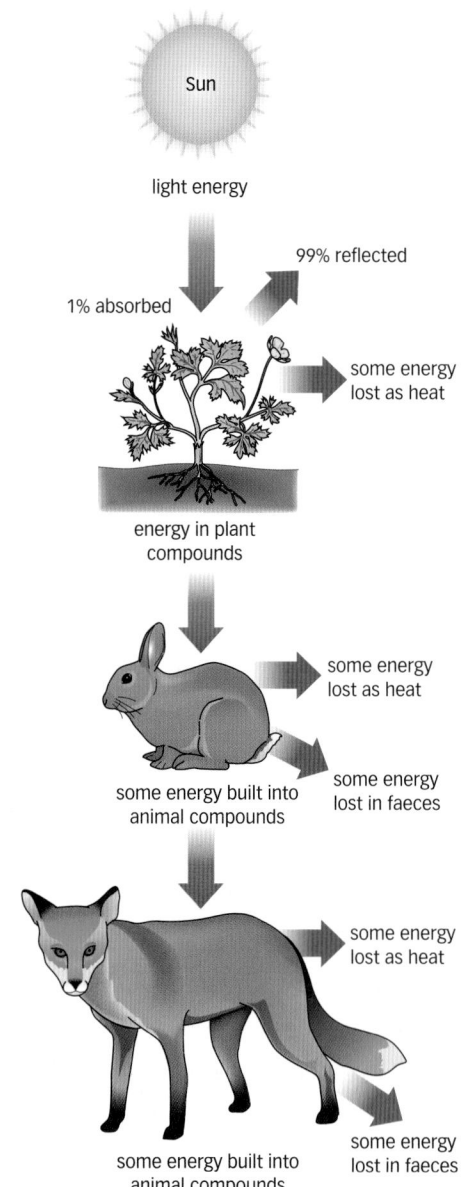

Sun

light energy

99% reflected

1% absorbed

some energy lost as heat

energy in plant compounds

some energy lost as heat

some energy built into animal compounds

some energy lost in faeces

some energy lost as heat

some energy built into animal compounds

some energy lost in faeces

▲ Energy flows through a food chain. Some energy is transferred out at each stage.

Energy

Food chains show not only the flow of materials in biomass from one organism to another, but also the flow of **energy**. All living things need energy to stay alive.

Energy enters the food chain when green plants or algae capture sunlight energy, converting it into chemical energy by means of photosynthesis. Energy flows through the food chain, and leaves it as heat or in waste materials. There must be a continual supply of energy into the food chain to maintain life. However, some of the energy leaves the food chain at each link.

How is energy transferred out of the food chain?

- Not all the food eaten by animals is digested. Some passes straight through the body undigested and comes out in droppings. This transfers energy out of the food chain.
- Every organism in the food chain uses some energy for respiration. This process releases energy for the animal's movement, and heat to warm the body. Birds and animals are 'warm blooded' – they keep their bodies warm at a constant core temperature. It takes a lot of energy from food to maintain this temperature, so they need to eat more food than cold-blooded animals.

Because energy leaves at each link in the food chain, there is very little energy left for the organisms at the end of the chain. This keeps food chains short. Most land food chains do not have more than five organisms. This is also why food pyramids tend to have fewer organisms and less biomass at the top, because there is less energy than at the base.

> **A** State three ways in which energy is transferred out of the food chain.
>
> **B** Name the source of energy for
> (a) a plant
> (b) a herbivore.

Energy efficiency in farming

Modern farming needs to produce food as cost-effectively as possible. Farmers need to minimise energy losses from the food chain to get the best yield, maximising energy **efficiency**. Science tells us the following, which helps the farmer:

- The shorter the food chain, the less energy is transferred out. Farm animals tend to be herbivores. This is more energy efficient. If farmers raised carnivores for us to eat, they would need to farm large numbers of herbivores to feed them.
- Less energy is transferred out from animals if they use less heat to keep themselves warm. In intensive farming, larger animals may be kept indoors in a barn, or smaller animals like chickens are often reared in cages. However, it is important to consider animal welfare issues; some battery conditions are inhumane.

Key words

energy, efficiency

Exam tip

✓ Remember that energy can't be created or destroyed. Never talk about energy being made or used up, simply say that it is transferred.

▲ Battery farming reduces energy losses

- Energy can also be kept in the food chain by reducing animals' movement. Again, this is achieved by keeping animals indoors or in cages.
- Farmers and growers need to minimise the energy that is transferred out to pests if they eat crops, or to weeds that compete with their plants for light.

Questions

1 Name the process by which light energy is converted into chemical energy by plants. ↓ E

2 Explain why fast-moving predators need a large amount of food. ↓ C

3 Explain why there are far more zebras than lions on the African savannah.

4 On free-range farms the animals are allowed to roam freely outdoors. Explain why this results in produce which is more expensive. ↓ A*

21: Recycling in nature

Learning objectives

After studying this topic, you should be able to:

✔ know that nature recycles by the decay of dead material

✔ know that **microbes** play an important part in the process of decay

Key words

microbe, decay

▲ A millipede eating leaf litter – a detritivore

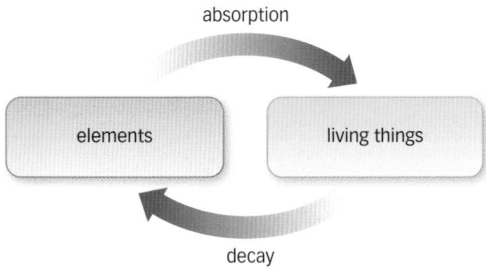

▲ Elements cycle between the living and non-living world

Did you know...?

Some of the carbon in your body is made up of the bodies of people in the past. Maybe some of it was part of King Henry VIII!

Round and round

Elements pass between the living world and the non-living world – air, water, soil, and rocks – in a constant cycle. Plants absorb elements, including carbon and nitrogen, and build them into useful molecules which help the plant grow. When an animal eats a plant, the plant's molecules become part of the animal.

Eventually, all plants and animals die. Their bodies **decay** and this decay process releases the elements back into the environment for plants to reuse. And so the cycle continues.

This is a kind of natural recycling process. Nature breaks down the remains of plants and animals to return the elements to the environment so they can be used again.

In a stable natural community like a woodland, this cycle keeps turning at a steady rate. The processes that remove materials from the environment and lock them up in plants and animals are balanced by the processes of decay, which return the materials to the environment.

▲ Fungi on dead wood – decomposers

Breaking down the dead

A number of organisms play a role in the process of decay. They break down dead plants and animals, and animal waste. There are two main groups of decay organisms:

- Detritivores, such as earthworms, maggots, millipedes, and woodlice, eat small parts of the dead material, which they digest and then release some as waste. This activity increases the surface area of the dead remains for decomposers to act on.
- Decomposers such as bacteria and fungi chemically break down dead material, releasing ammonia into the soil.

Saprotrophic feeding

Most decomposers feed by releasing enzymes on to the dead animal or plant. The enzymes digest the dead material, and the decomposers then absorb the digested chemicals. This process is called saprotrophic feeding.

Conditions for decay

Decay happens faster at certain times of the year, especially during autumn. This is because conditions are right for the bacteria and fungi:

- plenty of food (dead plants or fallen leaves)
- oxygen
- a suitable temperature
- moisture.

C Use your knowledge of microbes to suggest why the carbon cycle slows during the winter.

D Why is autumn a good season for the process of decay?

Organic gardening

Gardeners make use of natural recycling. They gather garden waste like grass cuttings, leaves, and twigs. They pile them up or put them into a compost bin, and allow them to decay. The result is a nutrient-rich compost which can be used to grow plants. Some local councils collect garden waste and put it in massive compost heaps. This reduces the use of landfill sites, and produces useful compost for parks and gardens.

A Name two important elements that plants need in order to grow.

B Describe why bacteria are important in natural recycling.

▲ Compost is a product of natural recycling

Questions

1 Name two types of organism that cause decay. ↓ E

2 Describe the difference between a decomposer and a detritivore. ↓ C

3 Describe how an organic farmer, who does not want to use manufactured fertilisers on his farm, could produce compost to help his crops grow. ↓ A*

Learning objectives

After studying this topic, you should be able to:

- ✔ know that elements are cycled between the living and non-living world
- ✔ understand the steps in the carbon cycle
- ✔ be aware of processes that cause an imbalance in the cycle

Key words

carbon cycle, photosynthesis, respiration

Carbon in the living and non-living world

Carbon is the one of the most common elements in living things. All of our major molecules, including carbohydrates, proteins, fats, and DNA, contain carbon. The process of carbon moving between the living and non-living world is called the **carbon cycle**.

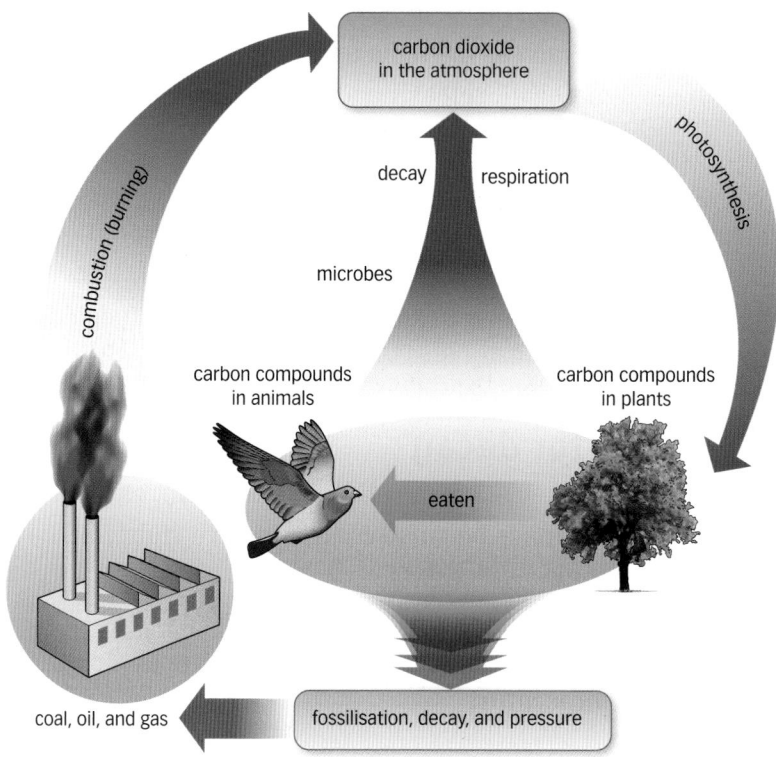

▲ The major steps in the carbon cycle

- Carbon is present in the atmosphere as carbon dioxide.
- Carbon dioxide is absorbed by green plants and algae and built into carbohydrate molecules such as sugars. This happens during **photosynthesis**.
- The plant uses some of the sugars to make other molecules such as cellulose, fats, and proteins, which it uses to grow.
- Plants are eaten by animals and so these carbon compounds can pass into animals and become part of their bodies.
- Both plants and animals **respire**. This returns some carbon dioxide back into the atmosphere.
- When plants and animals die, microbes digest their bodies in the process of decay and carbon dioxide is released back into the atmosphere.

- Not all plant and animal bodies will decay. Some are buried under layers of silt and over millions of years begin to fossilise.
- This forms fossil fuels such as coal, oil, and gas.
- Humans extract fossil fuels and burn them to release energy.
- Burning fossil fuels releases the carbon that was stored millions of years ago as carbon dioxide.
- In a stable environment, the amount of carbon dioxide released should approximately equal the amount absorbed.

Locked-up carbon

Carbon often gets built into the bodies of plants or animals and stays there for millions of years.

There are giant redwood trees in the forests in California which have been growing for more than 2000 years. They have built carbon into the molecules that make up the wood in the tree. Carbon has been locked up in the tree trunk for all that time. Even when these trees die, they are very difficult for microbes to break down. Detritivores are small animals that eat bits of dead plant remains, but they can only eat very small amounts of the dead redwood tree.

When humans cut down forests, the carbon locked up in the trees is released by decay, and there are fewer trees to absorb carbon dioxide from the air. Burning the trees releases the carbon more quickly. Deforestation upsets the balance of the carbon cycle.

A Humans burn vast amounts of fossil fuels. What is the impact of this activity on the carbon cycle?

B How might planting trees balance the carbon given out by your activities in daily life?

▲ Giant redwood trees in the forests of California contain locked-up carbon

Questions

1 (a) Name the process by which carbon enters living organisms.

   (b) Name two processes by which carbon is released from living organisms.

↓ E

2 Describe the key role of microbes in the carbon cycle.

↓ C

3 Describe how the actions of humans are leading to an imbalance in the carbon cycle.

↓ A*

Learning objectives

After studying this topic, you should be able to:

- ✔ state that most body cells contain chromosomes, which carry information in the form of genes
- ✔ state that genes control the characteristics of the body
- ✔ know that differences in characteristics may be due to differences in genes or the environment or both

▲ Everyone in the picture is human. However, within this large group there is a lot of variation. There are differences in the characteristics of different individuals.

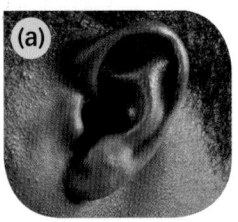 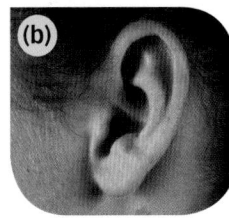

▲ The shape of your earlobe is determined by genes. (a) Attached earlobe; (b) free hanging earlobe.

Did you know...?

Chimpanzees have 24 pairs of chromosomes in each body cell and dogs have 39 pairs of chromosomes in each body cell.

Differences in characteristics

Although all humans are similar to each other, we are also different in many ways. We share characteristics like eye colour; we all have coloured eyes. However, some of us have brown eyes, some blue, some green, and so on. Different individual people show differences in their characteristics.

How human characteristics are determined

Some characteristics are determined by the environment, some by **genes**, and some by both genes and environment.

Characteristics determined by the environment

These include:

- scars
- learning to speak a language.

You get scars after you have injured yourself. Your children will not be born with the same scars.

At birth you cannot speak; you have to learn by listening to your parents and copying them. If you never hear anyone speak, you will not be able to develop language.

Characteristics determined by genes

These characteristics may be inherited. They include:

- eye colour
- earlobe shape
- nose shape.

Characteristics determined by both genes and environment

Some characteristics depend on both genes and suitable environmental factors, such as:

- intelligence
- height
- body mass.

For example, you may inherit genes which control your brain development in such a way that you will be able to have high intelligence. However, if you are not fed properly whilst growing up, or if no one talks to you or reads to you, and you are not given opportunities for stimulating play, you will not develop your full intelligence potential. You may have a genetic potential to be very tall, but you will not reach it if you are undernourished.

A State two human characteristics that are determined by the environment.

B State three human characteristics that are determined by genes.

C State three human characteristics that are determined by both genes and the environment.

Genes and chromosomes

Inside the nucleus of every cell there are thread-like structures called **chromosomes**. All of your body cells have the same number of chromosomes: 23 matching pairs of chromosomes. One member of each pair came from your father, in a sperm, and the other member of each pair came from your mother, in an egg.

Chromosomes contain genes, so half your genes came from your mother and half from your father. This is why you have some characteristics similar to each parent. Humans have 23 pairs of chromosomes, but organisms in different species have different numbers.

- Each chromosome contains many genes.
- Each gene has coded information that controls a particular characteristic.
- Different genes code for different characteristics of the body.

Each gene carries information in the form of coded instructions. These coded instructions control the activity of the cell and the characteristics of the organism.

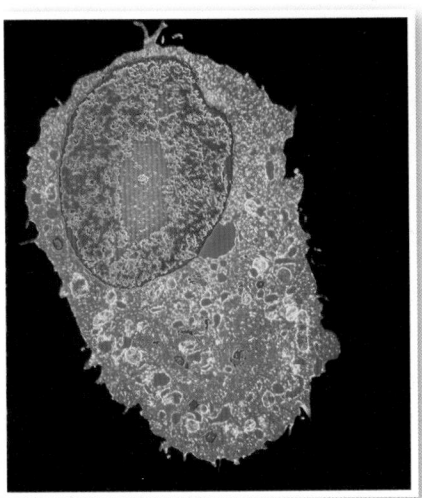

◄ A mammalian cell (×300). The cytoplasm is coloured blue and the large orange and green structure is the nucleus. Inside the nucleus are chromosomes containing genes.

Chromosomes. Some of the ► genes on some chromosomes have been tagged with a fluorescent chemical and show up as yellow on the red chromosomes.

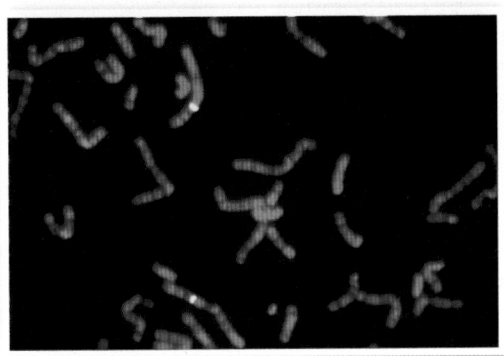

Key words

gene, chromosome

Exam tip **AQA**

✔ Remember that we have 23 pairs of chromosomes in our body cells. Our eggs and sperms have half that number, just 23. Remember also that other living organisms have different numbers of chromosomes from us.

Questions

1 Where, in animal and plant cells, are the chromosomes?

2 How many chromosomes are there in each of your brain cells?

3 What is a gene?

4 Where are genes found?

5 What do genes do?

6 Humans have about 20 000 genes. Why do you think we have so many genes?

7 Explain why differences in intelligence between individuals are not just due to genes.

↓ E

↓ C

↓ A*

Learning objectives

After studying this topic, you should be able to:

✔ know that there are two forms of reproduction – sexual and asexual

✔ recall that new plants can be produced quickly and cheaply by taking genetically identical cuttings from older plants

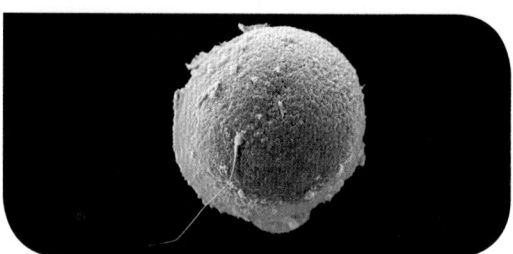

▲ Fertilisation: a human sperm penetrating a human egg (× 500)

▲ This hydra, a small aquatic animal related to jellyfish, can reproduce asexually. The small offspring branching out on the left will eventually break away and become independent. It is a clone of the single parent.

Sexual reproduction

Sexual reproduction involves two different **gametes** (sex cells) joining.

- Each gamete has only half the number of chromosomes of a normal body cell.
- The female gamete is called an egg. It has 23 chromosomes in its nucleus. It is larger than the male gametes. Adult female humans produce one egg each month.
- The male gametes are called sperms. Adult human males make hundreds of millions of sperms each day. Each sperm has a nucleus containing 23 chromosomes.
- One sperm will **fertilise** one egg.
- When this happens, the two nuclei fuse and the chromosomes from the father can pair up with chromosomes from the mother.
- The resulting new individual has chromosomes from two parents. They have a mixture of genetic information. This is why the offspring have genetic variation.

A What are gametes?

B How are male gametes different from female gametes?

C Explain why offspring are genetically different from each other and from their parents.

Asexual reproduction

Some organisms can reproduce **asexually**. Only one parent is needed, and

- there are no gametes
- there is no mixing of genetic information from two parents
- there is no genetic variation among the offspring
- the offspring are genetically identical to each other and to the parent.

Bacteria and some other single-celled organisms can reproduce asexually. Some more complex organisms, like the hydra shown on the left, can also reproduce asexually.

Cuttings

Some plants can reproduce asexually as well as sexually. We can take **cuttings** from plants. New plants produced by cuttings are all genetically identical to their single parent plant.

The advantages of taking cuttings are:
- The new plants are made quickly and cheaply.
- The new plants are genetically identical to the parent plant, and so will have the desirable characteristics of the parent plant.

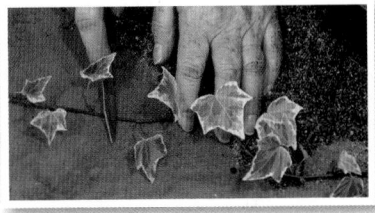

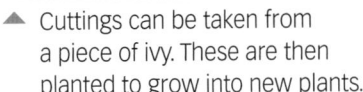

▲ Cuttings can be taken from a piece of ivy. These are then planted to grow into new plants.

Taking cuttings has been done for many years. The plants produced in this way are **clones**, because they are genetically identical to each other and to the parent.

Key words

sexual reproduction, gametes, fertilisation, asexual reproduction, cuttings, clones

Did you know...?

Identical twins are clones of each other.

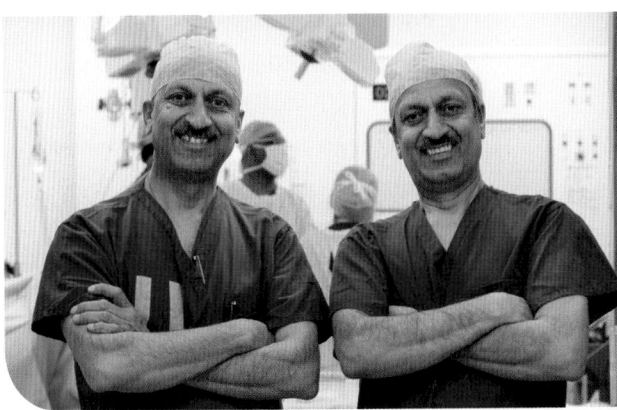

▲ Identical twin brothers. Not only do they look exactly alike, they are both senior operating assistants in a hospital.

Questions

1. State one way in which male gametes are the same as female gametes.
2. When a sperm fertilises an egg, what happens to the two nuclei?
3. Why are cuttings taken from a plant all genetically identical to the parent and to each other?
4. State two advantages of producing plants by taking cuttings.
5. Sometimes the new cutting is placed in a pot and a clear plastic bag placed over the whole thing. Why do you think this is done?
6. Plants reproduce sexually by making seeds. A plant grower has a variety of geranium plant that is particularly attractive and sells well. Should he use cuttings or seeds to grow lots of them? Explain your answer.

Exam tip   AQA

✔ When you are describing offspring produced by asexual reproduction, don't just say they are identical, say they are genetically identical to each other and to the parent.

▲ Geranium flowers

Learning objectives

After studying this topic, you should be able to:

✔ know that modern cloning techniques include tissue culture, embryo transplants, and adult cloning

✔ interpret information about cloning techniques

✔ make informed judgements about issues concerning cloning

Exam tip **AQA**

✔ Try not to be emotional if you are asked about uses of cloning. Consider the social and ethical issues and try to be objective.

A Why does the liquid or jelly used for tissue culture have to be very clean?

B Why are the calves produced by embryo transplant genetically identical to each other?

Did you know...?

Dolly the sheep was a celebrity and could not be allowed to run in the fields, as she was valuable. However, she gave birth to six lambs, all healthy and born naturally. She died, aged seven years old, of a type of lung cancer caused by a virus. The other (non-cloned) sheep kept in the same barn also died of it.

Tissue culture

Tissue culture can be used to make new plants by asexual reproduction. Scientists and technicians take small groups of cells from part of a plant and put them into a special liquid or jelly. The liquid or jelly has to be very clean so that there are no bacteria or moulds. It may also have some special chemicals to promote the development of these cells into root cells, stem cells, and leaf cells, so that new plants grow.

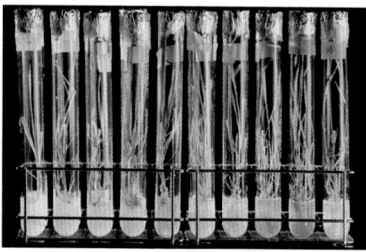

◀ Cereal plants being grown by tissue culture. All of these plants are genetically identical to the parent plant. They will all have the parent plant's desirable characteristics.

Embryo transplants

Embryo transplants may be used to produce cattle with desirable characteristics.

* Eggs are obtained from prized female cows and fertilised with a selected bull's sperm, in a dish.
* Each resulting embryo is allowed to develop in the dish to the eight-cell stage.
* Then it can be split into four two-cell embryos.
* These new embryos then each develop to eight cells.
* Each can then be put into the uterus (womb) of a less-valuable cow.
* These cows are host mothers (surrogates). The embryos develop inside them and the calves are born.
* The calves are all genetically identical to each other.

In this way, one valuable cow with good characteristics can have many offspring within a short space of time.

 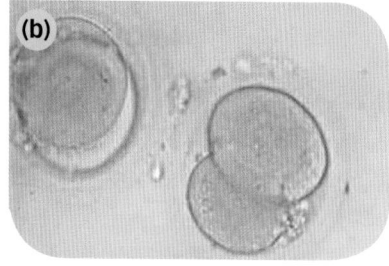

▲ (a) Cow embryos being removed from storage in liquid nitrogen
(b) Embryos at the two-cell stage

Adult cell cloning

Adult cell cloning was first carried out in 1996 in Scotland by Professor Ian Wilmut and his team.

- An unfertilised egg was taken from a Scottish Blackface ewe (female sheep).
- Its nucleus was destroyed.
- A cell was taken from the mammary gland of a six-year-old Finn Dorset ewe.
- Its nucleus was implanted into the empty Scottish Blackface egg.
- An electric shock was given to the resulting egg cell to make it divide.
- It developed into an embryo that was put into a host mother sheep.

The resulting cloned sheep was Dolly, a Finn Dorset ewe.

◀ Professor Ian Wilmut and Dolly

Dolly was produced because scientists had created genetically engineered sheep that can make useful medicines for humans in their milk. These sheep could breed, but their offspring may not inherit the human gene that made their milk useful. Also, half of their offspring would be male and would not make milk. If these sheep could be cloned instead, then many sheep able to make the valuable medicine could be created.

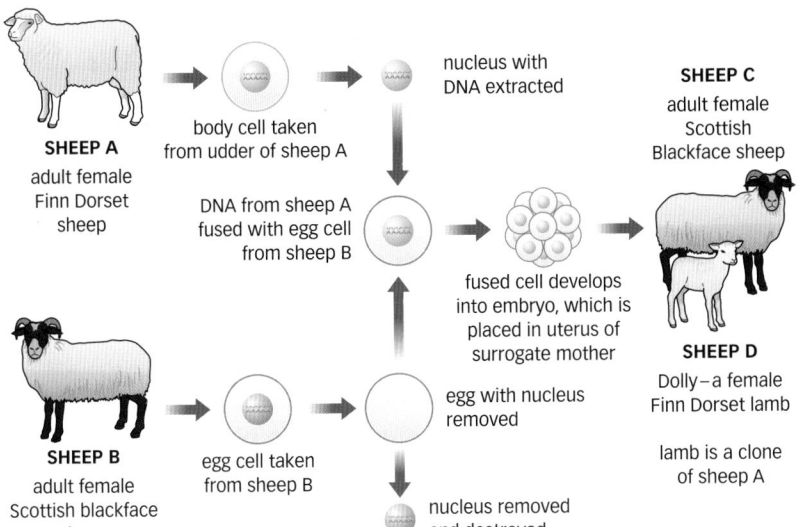

SHEEP A
adult female
Finn Dorset
sheep

body cell taken
from udder of sheep A

nucleus with
DNA extracted

DNA from sheep A
fused with egg cell
from sheep B

fused cell develops
into embryo, which is
placed in uterus of
surrogate mother

SHEEP C
adult female
Scottish
Blackface sheep

SHEEP B
adult female
Scottish blackface
sheep

egg cell taken
from sheep B

egg with nucleus
removed

nucleus removed
and destroyed

SHEEP D
Dolly – a female
Finn Dorset lamb

lamb is a clone
of sheep A

▲ Flow diagram showing adult cell cloning

Key words

tissue culture, embryo transplant, adult cell cloning

Questions

1 What is adult cell cloning?

2 Why are the calves produced by embryo transplant not genetically identical to
(a) their true mother?
(b) their surrogate mother?

3 Explain how Dolly, a Finn Dorset ewe, was born to a Scottish Blackface ewe.

4 Dolly's birth was kept secret for six months whilst the researchers wrote their paper and had it peer-reviewed before publication. When the news broke in 1997, some of the headlines included: '*Golly Dolly! It's the abolition of Man.*' '*Terrified researcher tells of how Dolly kills and eats a lamb.*' '*The clone rangers need to be stopped.*' '*Human cloning not far away.*' Comment on the use of such sensationalism.

5 Human cloning is illegal and is not carried out (except in nature in the form of identical twins!). Discuss the social and ethical issues around animal cloning.

6 The press reported that Dolly died young because she was cloned. Does the information about her death on page 68 support their view?

Learning objectives

After studying this topic, you should be able to:

✔ state that in genetic engineering genes can be transferred from one cell to another cell

✔ state that genes can also be transferred to the cells of animals or plants at an early stage in their development, so that they develop the desired characteristics

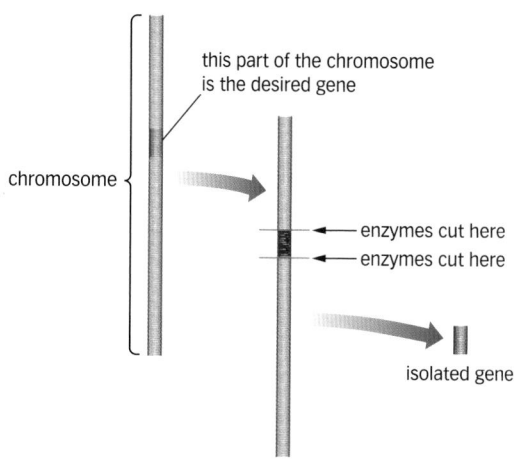

▲ Using an enzyme to cut out a gene

A How can scientists cut out genes from human chromosomes?

B What are the advantages of using genetically modified bacteria to make human insulin?

What is genetic engineering?

Genetic engineering means changing an organism's genes.

In the past, selective breeding programmes have been used to introduce genes from one variety of plant or animal into another. This takes a long time – around 20 years of careful breeding.

Now scientists can make use of various **enzymes** to manipulate genes, and to speed up the process of transferring genes from one organism to another.

Cutting out genes

Chromosomes and genes are made of **DNA**. Scientists have obtained special enzymes from bacteria, and these enzymes can cut DNA at particular places. There are many different types of these enzymes, and different ones can be used to cut DNA. If a gene is cut at either end, it can be cut out from its chromosome.

Making insulin: genetically modified bacteria

Genetic engineering is used to produce human insulin. The gene for human insulin is put into bacteria. The bacteria then multiply and produce insulin. The insulin is harvested from the bacteria and used to treat people with diabetes. This method is better than the old method of obtaining insulin from pigs' pancreases because

- The bacteria can be easily grown in large amounts.
- Scientists can now make enough insulin to treat all the people with diabetes.
- The insulin the bacteria make can be easily purified.
- It is suitable for vegetarians or anyone who objects to obtaining products from pigs.
- There is no risk of transmitting a disease.

Treating emphysema: genetically modified sheep

Transferring genes to early embryos

Human genes that have been 'cut out' can be put into cells of another organism, such as a mammal or a plant. The genes need to be transferred into an early embryo. Then, as the embryo divides, all the cells of the resulting individual will have the new gene.

Genetically engineered female sheep have been produced that have a gene for a human protein. This protein is too large to be made in bacterial cells. The genetically modified sheep make the human protein in their milk. The sheep can be milked and the human protein is extracted from the milk. It is used to treat people with hereditary emphysema.

The genetically engineered sheep are produced in this way:

> The human gene for the protein is 'cut out' from a human chromosome.

↓

> Several copies of the gene are obtained.

↓

> They are put into fertilised sheep eggs, in a glass dish.

↓

> The fertilised eggs are allowed to divide into a ball of cells.

↓

> One cell is taken from each ball of cells, and the chromosomes are checked to find out which ones are female.

↓

> The balls of cells that will develop into female sheep are put back into the female sheep wombs.

↓

> The embryos develop and lambs are born.

The advantages of producing the human protein in this way are
- The sheep are not harmed in any way.
- The sheep are kept on farms in good conditions.
- The protein is only made in their milk.
- The protein can be easily collected without hurting the sheep – just as we milk cows.
- These sheep are valuable and so will have a longer life than many other sheep that are killed for meat.
- The protein can be easily separated from the milk, and is pure and uncontaminated.
- A lot of the protein can be made in order to treat people who are ill.

One disadvantage is that this can only be done in parts of the world where sheep can live. However, the medicine produced can be exported all over the world.

Key words

genetic engineering, enzyme, DNA

Exam tip **AQA**

- ✔ Always remain objective when discussing ethical issues. Try to give a balanced argument.

Questions

1. Why are genetically modified sheep used to make a human protein to treat hereditary emphysema?

2. Why is the human gene put into an early embryo sheep and not into a developed lamb?

3. Why do you think only female embryos are used?

4. Discuss the ethical issues of making the human protein in this way.

Learning objectives

After studying this topic, you should be able to:

- ✔ understand that new genes can be transferred to crop plants, which are then called genetically modified crops
- ✔ know that some GM crops are resistant to insects or to **herbicides**
- ✔ know that GM crops may show increased yields or improved nutritional content

Key words

genetic modification, herbicide

Did you know...?

If potatoes and tomatoes (both of which originated in South America) were introduced to us today as novel foods, they might well not get past safety tests. People would probably object to their use as foods, as parts of them contain toxins. They are both in the deadly nightshade family.

Genetically modified (GM) plants

Humans have been selectively breeding plants for about 10 000 years. They have always tried to increase the yield or make a crop grow in an environment where it does not normally grow. **Genetic modification** (GM) speeds up the process.

The human population has increased a lot and is still growing. By 2030 it will be about nine billion. More people, and their roads and buildings, take up more space. So we have to grow more food on less land. It is very unlikely that we can do that without using GM plants.

Using GM crops will not be the only way to improve food production, but it is likely to be an important one. We cannot just keep on using more fertiliser and pesticides, as they have adverse effects on the soil and environment.

Examples of GM crops

GM crop plant	Use
Soya beans	Resistant to weedkiller, so that weeds can be sprayed and killed without killing the soya. Some 77% of all soya grown in the world is genetically modified. Most is grown in the US and Brazil.
Corn (maize)	Resistant to a pest, the corn borer. This is a tiny insect that eats into the corn stalk so that the plant falls over and does not produce seeds for humans or livestock to eat. The holes also let fungi get in, and these produce toxins. GM maize has been grown in the US and Canada since 1997. Some 80% of the maize grown in the US is GM. GM maize is now also grown in Spain, Portugal, the Czech Republic, and Germany.
Golden rice	Contains vitamin A in its grains. White rice does not contain vitamin A. Rice is the major part of the diet in many developing countries. Each year 600 000 children worldwide go blind due to lack of vitamin A. Many of these children die within a year of going blind, as vitamin A is also needed for growth and protection from infections. Golden rice contains vitamin A and is a good way of providing enough vitamin A to children in developing countries at no extra cost.

Cotton 	Resistant to pests. Cotton fibres are used for textiles, and the seeds provide oil and protein for animal feed or oil for margarine. The cotton seed capsule (boll) is attacked by caterpillars. The plants used to be sprayed with chemical pesticides but often these did not kill the caterpillars. The chemicals stayed on the outside of the boll and the caterpillars are inside. GM cotton has a gene from a bacterium. The gene codes for a toxin that kills the caterpillars. This Bt toxin has been used for decades by extracting it from bacteria. Now with GM cotton, the cotton plants themselves make the toxin and kill the caterpillars even if they are inside the boll. Most GM cotton is grown in India and the US, but some is grown in Argentina, Mexico, South Africa, Australia, and Columbia. 68% of cotton grown in China is GM.
Tomatoes, potatoes, squash, papaya 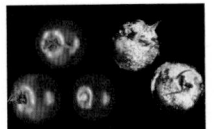	Resistant to pests. If crops are resistant to pests, fewer chemicals have to be used. This reduces the risk of killing useful insects or of the chemicals entering the food chain. Three of the tomatoes on the left are GM. They are resistant to a mould fungus.
Bananas 	Resistant to pests and containing extra nutrients, like zinc. In some African countries bananas form a great part of the diet. Zinc is an important mineral, and where people do not eat much meat, they often do not get enough zinc. Farmers can grow GM bananas that contain more zinc and are resistant to diseases. They do not need to use pesticides. The farmers are not exposed to the harmful chemicals, and neither is the environment.
Corn	Makes 'fish oils'. These oils are normally obtained from oily fish and help human brain development. However, fish do not make these oils. They get them from eating algae. The gene from algae has been put into a variety of corn. This could provide the 'fish oils' humans need, particularly as fish stocks are dwindling.
Tomatoes, melon	Longer shelf life, as ripening is delayed. This means the fruits can be ripened on the plants before being picked. This enables them to have more flavour.

Questions

1. Why do humans need to increase the amount of food grown in the world during the next 20 years and beyond?

2. Humans have increased food production greatly over the last 50 years. They grew new varieties of crop plant that had better yields. How were these varieties produced?

3. The 'green revolution' that started in the 1960s involved using much more pesticide and fertiliser chemicals. Why can we not just increase the use of these chemicals over the next 20 years to increase crop yields?

4. Were there any harmful effects from these chemicals?

5. Explain why using GM crops can reduce the amount of fertiliser and pesticide chemicals used.

6. What is the advantage of golden rice, compared to normal white rice?

7. Why may GM corn that contains 'fish oils' be important for humans in the future?

Learning objective

After studying this topic, you should be able to:

✔ make informed judgements about the economical, social, and ethical issues surrounding GM crops

Key words

hectare, fertiliser, pesticide, yield

A researcher at a research station in the UK compares growth of GM crops with non-GM crops

GM crops are controversial

Trials have to be carried out on new GM crops to see if there are any associated health risks or risks to the environment. Protesters have sometimes destroyed the trials, so they have to be carried out in secret places. Because they are controversial, many GM crops developed nearly 20 years ago are still not being grown, although they are safe and could help many people.

Genetic engineering was used to develop GM crops because it was faster than using a selective breeding programme. However, the 'red tape' of bureaucracy, together with public pressure, has held up the commercial growing of many of these crops. Some, such as golden rice, were developed 20 years ago and are still not being grown. Had this rice been developed using selective breeding it would also have taken 20 years and would have been introduced by now without protest. It would have saved many lives.

What are the objections?

When Sainsbury's first introduced tomato purée made from GM tomatoes in 1991, it was clearly labelled and people had no objections. It tasted good and sold well.

▲ Tomato puree made from GM tomatoes

Monsanto, a large US-based company, has developed strains of GM corn and soya. At first they did not label their GM produce, and this upset many people, as it took away their ability to choose. One supermarket chain reacted by saying they would not sell any GM produce. Other supermarkets followed suit in order not to lose trade, and the press joined in with sensationalised articles about 'Frankenfoods'.

We need to look objectively at some of the arguments.

- Many people oppose the idea of eating GM crops. Some think it is 'interfering with nature'. However, this is what humans regularly do – vaccination, surgery, curing illnesses, and growing more food per **hectare** (by using **fertilisers** and **pesticides**) to feed a growing population are all interfering with nature. You could argue that farming is interfering with nature. However, humans are also part of nature, and we harness it to meet our needs.

- Some trials show that GM crops do not produce a higher **yield**. Other trials show that they do. More trials need to be carried out. In some cases the purpose of genetic modification is to improve the nutrition or to make the plants resistant to disease. So individual plants will not produce more yield, but each hectare will if pests are not destroying crops.

- Some GM crops are resistant to pests, so they need fewer pesticides. This is good for the environment, as the pesticides often kill useful insects as well as harmful ones. A particular pesticide has recently been banned in Germany and France because many dead bees have been found to have residues of it in their tissues. However, this pesticide has not yet been banned in the UK.

- In 2003, Zambia banned the import of GM crops and this led to widespread famine. In 2005, they allowed GM maize in.

- Most people who oppose GM crops live in developed countries and are not at risk of starvation or of becoming blind due to lack of vitamin A.

- GM foods have been eaten in the US for well over 10 years and have not caused health problems. This is a good natural experiment – the US population compared to the control group in Europe.

- Golden rice had to be trialled to see if it caused allergies now that it contained vitamin A, even though humans have been eating carrots (a good natural source of vitamin A) for thousands of years. Much of the non-GM food we eat has never been tested.

- Many non-GM foods (such as processed foods) do cause health problems because they contain a lot of saturated fat and salt. Both of these are clearly linked to heart disease and to cancer (35% of all cancers are diet-related).

- Studies of areas where GM crops are grown, compared with areas where they are not grown have shown that GM crops do not appear to affect the local wildlife, neither plants nor animals. However, many other human activities do affect wildlife, such as extracting and burning oil.

Did you know...?

At the time of writing there are no commercially-grown GM crops in the UK. A few are grown in Spain, Portugal, and Germany. Most are grown in the US, Canada, Brazil, China, India, and Australia.

Questions

1 Why was genetic engineering used to produce new varieties of crops?

2 Make a table. On one side list arguments in favour of GM crops. On the other list arguments against GM crops. Share your ideas with the rest of the class. Make a poster summarising the class's pros and cons.

3 Use the Internet to find out more about:
   (a) golden rice
   (b) Flavr Savr tomatoes
   (c) GM soya
   (d) GM potatoes
   (e) GM maize.

You might like to work in small groups, each group researching one of the topics. Then prepare a presentation to share your information with the rest of the class.

Learning objectives

After studying this topic, you should be able to:

✔ know that in classification, organisms are grouped based on similarities and differences

✔ know the characteristics of some of the major groups

Key words

classification, kingdom

A List features that are common to both cats and dogs, ie characteristics used to place them in closely related classification groups.

B List features of goldfish and crabs that they don't share, ie characteristics used to place them in different classification groups.

C Why do biologists classify living things?

D New organisms are discovered every day. How would you go about deciding which group a new discovery should be placed in?

Similarities and differences

The world is full of millions of different types of living things. Biologists put living things into groups, which makes them easier to study. This grouping process is called **classification**.

To classify living things biologists observe their features (characteristics). You can observe both similarities and differences in these two types of daffodil.

◄ Similarities:
- same-shaped leaves
- same number of petals

Differences:
- different colour
- different height

Living things that share lots of similar characteristics are grouped together. If organisms have lots of differences in their characteristics then they are classified in different groups.

Kingdoms

All living things are placed into major groups called **kingdoms**. Each kingdom has different characteristics.

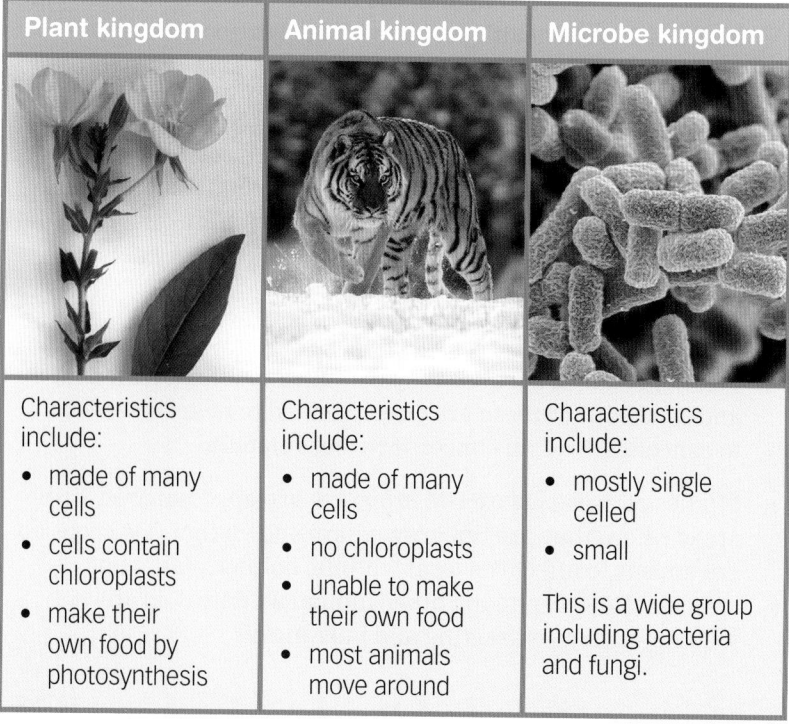

Plant kingdom	Animal kingdom	Microbe kingdom
Characteristics include: • made of many cells • cells contain chloroplasts • make their own food by photosynthesis	Characteristics include: • made of many cells • no chloroplasts • unable to make their own food • most animals move around	Characteristics include: • mostly single celled • small This is a wide group including bacteria and fungi.

Evolutionary links between species

If two species are closely related, then they tend to share more characteristics. This is because they both had the same close ancestor. Chimpanzees and humans share a number of features. They both

- have a backbone
- have skin covered in hair
- feed their young with milk
- have grasping hands.

This is because they both had the same common ancestor. If you compare these two species with a lizard, there are far fewer similarities, they all have a backbone, for example.

Variations to suit a habitat

Closely related species may look very different if they live in different habitats. The Arctic fox lives in an arctic habitat and the fennec fox lives in a desert habitat. They still share characteristics.

On the other hand, organisms that are not closely related may share several features if they live in the same habitat. For example, the shark and the dolphin both have an ocean habitat. They show some similarities but they are not closely related.

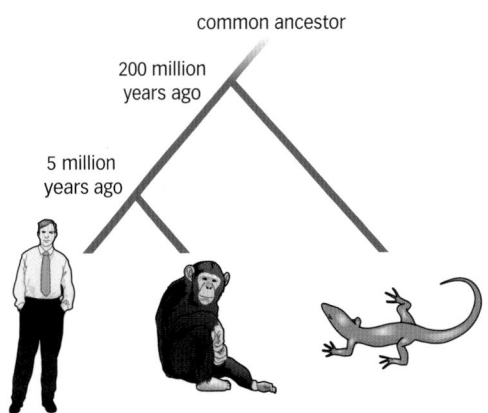

▲ This family tree shows the relationship between three species. The human and chimp are far more closely related to each other than they are to the lizard.

Did you know...?

The ancient Greeks were the first to try to organise and describe living things. Aristotle attempted to group animals, whilst his student and friend Theophrastus did the same for plants.

Exam tip AQA

✓ When classifying organisms, list their similarities and differences.

▲ Arctic fox

▲ Fennec fox

Questions

1 State the main features of a plant.

2 Explain why humans are classified as animals.

3 Sketch a possible family tree for the following four organisms, based on their similarities: zebra, horse, lizard, and goldfish.

E ↓ C ↓ A* ↓

Learning objectives

After studying this topic, you should be able to:

- ✔ know the causes and effects of evolution

Key words

evolution, natural selection, mutation

A Define evolution.

B What is an ancestor in the theory of evolution?

Evolution in action

Evolution

The world is full of millions of different species. Where did they all come from? This has puzzled biologists for many years. One big question is that of how life first began. Biologists now believe that simple life forms evolved in primitive oceans from large complex molecules. From these simple beginnings, all other organisms have evolved. This has taken over three billion years.

Other questions about where all the different organisms came from include:

- Why are there so many different species?
- How do species change or adapt over time?
- How did all the different species form?
- Why are some species closely related to each other?

Most biologists believe that an idea known as evolution best answers all four questions.

There are many different habitats in the world. In each habitat there are organisms that are well adapted to survive. This results in lots of species. But the habitats of the world are constantly changing. The organisms must also change to survive.

Evolution is the gradual change of an organism over time. This idea suggests that gradually one type of organism, called the ancestor, might change over many generations into one or more different species. This generates lots of different species over time. But how do organisms change?

Natural selection

The biologist Charles Darwin suggested an idea to explain how one species can change into another. This is now widely accepted by most biologists and is called **natural selection**. This theory can be used to explain how the giraffe's long neck evolved. There are four major steps in the theory.

1. Large populations showing variation

- Most species produce lots of offspring. This should cause a massive population growth for every species. One original pair of giraffes would produce millions of giraffes over a few hundred years.
- Individuals in a population will show a wide range of variation, because of differences in their genes.
- This variation is caused by chance **mutations** in genes.

2. Survival

- Not all of the organisms in a population survive and reproduce.
- Their survival is affected by changes in the environment. Some die from disease. Some starve. Some are eaten by predators. Some cannot find a mate.

3. The fittest

- Some of the variations will be an advantage. For example, some giraffes have longer necks than others.
- Longer necks allow those giraffes to reach leaves higher on the tree.
- When food lower down is scarce, those giraffes without the advantage of a long neck will die.

4. Passing on the advantage

- The surviving giraffes are the only ones that reproduce.
- Their offspring inherit the advantage. The gene for long necks has been passed on.
- Over many generations the number of giraffes with long necks increases.
- The result is that the giraffe species has evolved a long neck.

Questions

1 Darwin used the term 'survival of the fittest'. Explain what this means.

2 Pet shops sell white and brown rabbits. White rabbits are easily seen by foxes. Use Darwin's theory of evolution to explain why white rabbits are rare in the wild.

3 Explain why evolution does not often happen over short time periods.

E
↓
C

A*

A Explain why Darwin could not suggest how a new species forms.

B Explain why mutations are central to the formation of new species.

Forming new species

Sometimes two separate groups of the same **species** evolve differently. Each group gradually changes over time, becoming more different from each other. This can result in two new species.

Darwin did not suggest a mechanism for the formation of two new species, because at the time people did not understand inheritance and variation. Modern biologists have explained the process of forming new species:

- A single population becomes separated into two groups. The environment may be different for the two groups.
- Over time, each separated group of the population evolves differently.
- The longer they are separated, the more different they become.
- Eventually the two sub-populations have changed so much that they can no longer interbreed.
- They have formed separate but closely related species.

▲ The African elephant (left) and the Asian elephant (above) are closely related species, which both evolved from a common ancestor many years ago when they were separated by great distances.

Variation is the key to the development of new species. Mutations of genes lead to inherited variation in characteristics. Where useful mutations occur, they may cause a rapid change in the species.

Evidence for evolution

Biologists have looked for evidence that evolution happens by natural selection and survival of the fittest. They have found scientific observations that support the idea of evolution.

The peppered moth (*Biston betularia*)

Perhaps the best known example of evolution in action involves a moth called the peppered moth. This moth is pale in colour, and looks as though it has been sprinkled with pepper.

The pale moth was well camouflaged against the light bark of trees. During the 1800s, trees in industrial areas became covered in soot particles, and the bark became much darker. The moths were no longer camouflaged, and they were eaten by birds.

A mutation occurred in some moths, making them much darker in colour. The darker moths now had the advantage of camouflage.

▲ Light and dark varieties of peppered moth (*Biston betularia*)

Over the next 50 years, the dark variety became more common.

Today, in cleaner areas the light form of the moth is more common again. In industrial areas the dark form is still the more common form.

Key words

species

Questions

1 Use Darwin's theory of natural selection to explain the steps involved as the moths changed to the darker form.  **E**

2 What happens to the dark-coloured moths in the cleaner areas of Britain?

3 Lions and tigers are two closely related species which evolved from a common ancestor. Explain how moving to different habitats has resulted in the formation of these two different species. **C**

4 When explorers discovered Australia they were amazed by how different the animal species were from those back home. Explain why Australian animals are so different from European ones. **A***

Learning objectives

After studying this topic, you should be able to:

✓ know how Darwin collected his evidence

✓ evaluate other theories of evolution

▲ Charles Darwin aged 40

▲ HMS *Beagle*

How did Darwin gather his evidence?

Charles Darwin took years to come up with his theory of evolution. He had studied living things at university, and became the ship's naturalist on the HMS *Beagle*. During his voyage he collected evidence which helped him develop his ideas on evolution.

The ship visited the Galapagos Islands off the coast of South America. Darwin visited each of the islands and made notes on the different types of plants and animals. Darwin was struck by the variety of types of finches he found on the islands, which he thought were unrelated species. But after studying them, he realised that they were closely related species.

Darwin developed his ideas of evolution to explain how the different species of finch might have arisen.

- An ancestral finch species arrived on all the islands.
- Gradually over time the finches on separate islands changed independently.
- Some birds became adapted to feed on insects.
- Other birds adapted to eat seeds or fruit.
- This gave rise to lots of different finch species, that were different on the different islands.

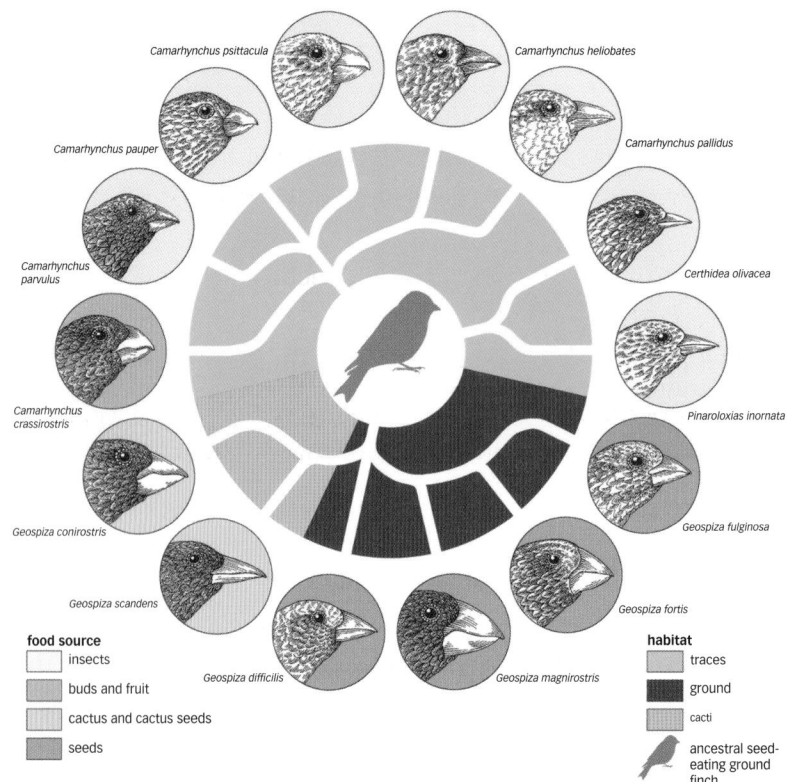

▲ These Galapagos finches all evolved from a common ancestor

How was Darwin's theory accepted?

Charles Darwin published his ideas in a book called *On the Origin of Species by Means of Natural Selection* in 1859. Many people were horrified by his ideas. The main objections were:

- Most religions have an account of how life started – they describe humans and other species being created. Darwin's theory seemed to disagree with the religious viewpoint. People particularly did not like the idea that humans could have evolved from apes.
- There was not much evidence at that time to convince other scientists that species evolve from each other.
- Scientists then did not know about genes, and so could not explain how characteristics could be passed on to allow inherited variation and evolution.

> **A** Scientists work by making observations. What sort of observations would Darwin have made of the finches to be able to explain their evolution?

An alternative idea: the work of Lamarck

Jean Lamarck was a French biologist who worked before Darwin. He suggested a different way to explain how species are formed in evolution. He would have explained the evolution of the giraffe as follows:

In a giraffe population, some giraffes want to feed off the leaves high on a tree. They have an 'inner need' to stretch their necks.

- The giraffes stretch their necks, becoming more successful.
- They pass their longer necks on to their offspring.

The ideas put forward by Lamarck are not accepted today. Living things cannot decide to alter their body because of an 'inner need'. Their bodies might change during their lifetime – for example, someone who works out in the gym will develop bigger muscles. However, this characteristic will not be passed on to their children. The only changes in organisms that can be passed on to offspring are changes in genes, not changes acquired during the organism's lifetime.

Exam tip  **AQA**

✔ When comparing the ideas of two different scientists, list the key points of each one and say what is similar and what is different.

> **B** Explain why it was important that Darwin wrote his book and spoke at scientific lectures to tell people about his ideas.

▲ Jean Baptiste Lamarck

Questions

1 Give three clear reasons why Darwin's theory was not immediately accepted. E

2 Explain why modern biologists think that Lamarck's ideas are not correct. C

3 The ideas of both Darwin and Lamarck can be used to explain the evolution of the giraffe's long neck. Explain how each of these scientists would have accounted for the giraffe's long neck. A*

Course catch-up

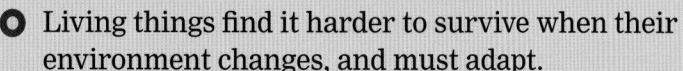

Revision checklist

- ⭕ Living things find it harder to survive when their environment changes, and must adapt.
- ⭕ Indicator species show how polluted air or water is. Sulfur dioxide levels affect lichen distribution, and dissolved oxygen concentrations control invertebrate distributions.
- ⭕ The mass of living material (biomass) gets less as you move along food chains or up pyramids of biomass.
- ⭕ Energy flows from the plants that capture it to the animals that make up the rest of the food chain, but some is lost at each step as heat and in waste materials.
- ⭕ The elements in living things are recycled by detritivores, like worms, and decomposers, like fungi and bacteria.
- ⭕ Photosynthesis traps carbon from the atmosphere in biomass. It passes along food chains and returns to the atmosphere when living things respire or decompose.
- ⭕ Our characteristics are controlled by the genes we inherit, and by environmental factors like our nutrient intakes.
- ⭕ Sexual reproduction fuses sex cells to produce offspring with genetic variation.
- ⭕ Asexual reproduction gives identical clones from one parent.
- ⭕ Clones are produced from plant cells, separated embryonic stem cells, or egg cells controlled by body cell nuclei.
- ⭕ Genetic engineering transfers genes to other organisms. If early embryos are used, they develop the characteristics the genes code for.
- ⭕ GM crops can resist insects or herbicides and produce increased yields or more nutritious crops.
- ⭕ Some people worry that GM crops could harm human health or spread their genes to wild flowers and insects.
- ⭕ To show the relationships between living things, we use their similarities and differences to classify them into groups.
- ⭕ Darwin's theory of evolution by natural selection is the most widely accepted explanation of evolution.
- ⭕ New species can form when populations are divided between different environments.
- ⭕ Darwin's theory took years to be accepted because it conflicted with religious accounts, there was limited evidence, and no-one could explain inheritance.

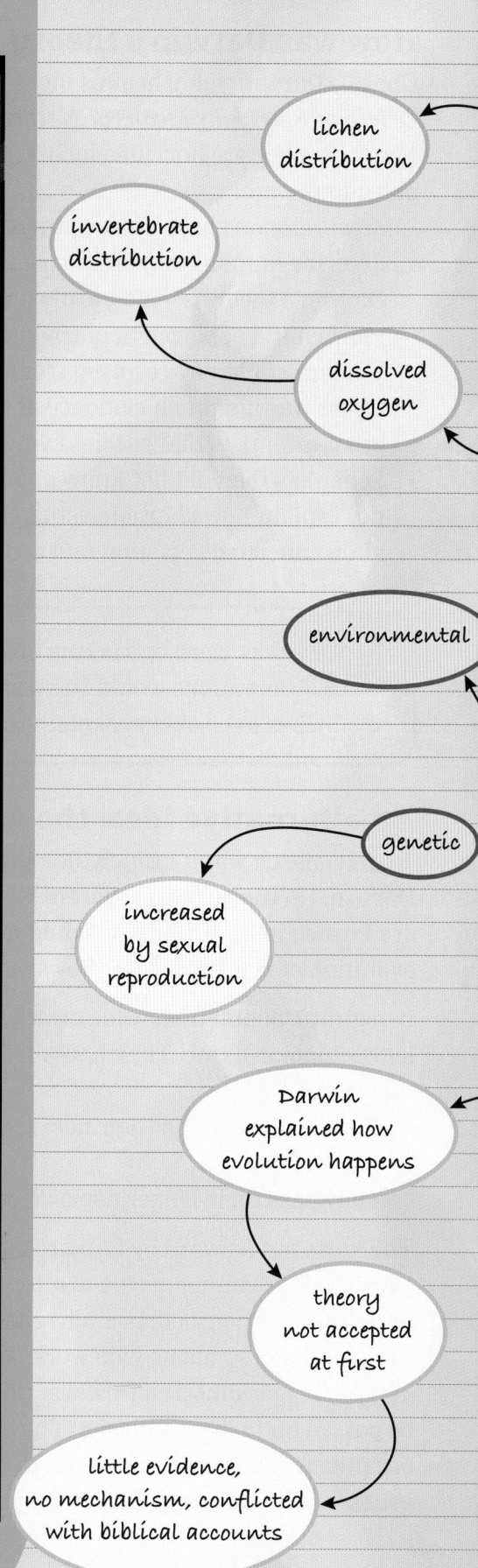

so their biomass gets less

each organism loses energy as heat and in waste

organisms must adapt or move away

bacteria and fungi

sulfur dioxide

make survival harder

decomposers

flows along food chain

highlight pollution

energy

changing environments

recycled

indicator species

elements

detritivores

energy and resources

eg worms

eg carbon cycle

SUMMARY

photosynthesis traps carbon in biomass

variation

adult cell cloning

similarities and differences used for classification

reduced by asexual reproduction

food chains transfer it to animals

new species arise when populations adapt to different environments

genetic engineering

tissue culture

embryo splitting

respiration and decay return it to the air

transfers new genes to organisms

makes GM crops

makes embryos develop the characteristics the genes code for

higher yields more nutrients

provoke concerns about health and environment

AQA Upgrade

Answering Extended Writing questions

Genetically modified organisms (GMOs) can be created that produce medicines for humans.

What are the benefits and risks of producing medicines in this way? Explain why some people are concerned about the use of GMOs.

The quality of written communication will be assessed in your answer to this question.

G–E

Some animals different geans. People say this is against nature. However people used to say that about IVF but the man who did it got the No Bell prize and millions have now got kids.
If I had an illness Id want to be cured and woudnt care where the medicine came from.

Examiner: This candidate knows what genetic engineering is. No specific examples are mentioned, such as bacteria making insulin or sheep making medicine to treat emphysema. The candidate mentions one concern that some people have and tries to give a counterargument, but goes off the point. Quite a lot of grammatical errors.

D–C

Some sheep have been engineered to make medicine. Some people think this is playing god and interfering with nature. But if they had emphysemia they would want to be cured. The medicine is in the sheeps milk so they are not harmed, just milked and wont be killed to eat.
Genetic engineering means changing your genes.

Examiner: This starts well with a good example. It gives some advantages and describes the concerns of some people. It explains what genetic engineering is, but the answer is not very well organised. There are a few grammatical and spelling errors. This answer is in the 3–4 mark band.

B–A*

Genetic engineering means changing an organism's genes. Bacteria can be engineered to make insulin for diabetics. This is better than getting it from pigs' pancreases as vegetarians won't object. Lots of bacteria can be grown in labs anywhere to make lots of insulin.
Some people think that scientists should not change an organism's genes as this interferes with nature. They may worry that the bacteria could escape from labs and make people ill.

Examiner: This is a well expressed answer. It is well organised and explains what genetic engineering is, covering some benefits and possible risks of genetic engineering. The spelling and punctuation are good. Some of the concerns that some people have are described.

Exam-style questions

1 Match these ways of producing offspring to their uses.

A02

Reproduction method	Use
sexual reproduction	growing identical plants
tissue culture	making a cat with the same genes as its mother
embryo splitting	growing plants with genetic variation
adult cell cloning	producing identical farm animals

2 Jen measured oxygen concentrations in a stream that runs past a pig farm.

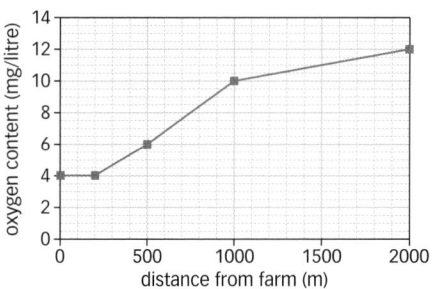

A03 **a** Where was the oxygen content of the water lowest?

A03 **b** How much did it rise between 500 m and 1000 m from the farm?

Jen took water samples and counted the number of mayfly nymphs in them.

Distance from farm (m)	Mayfly nymphs
0	0
200	0
500	6
1000	34
2000	35

A03 **c** What can Jen conclude about mayfly nymphs?

A02 **d** Jen has a hypothesis. She thinks that the pig farm is polluting the stream. Suggest other measurements she could take to strengthen her evidence.

3 The UK public has a history of opposition to GM crops. As at 2010, no GM crops had been grown in the UK for sale.

A02 **a** Suggest a useful characteristic that a GM crop could have.

A02 **b** What concerns might people have about growing GM crops in this country?

A02 **c** What concerns might people have about eating GM foods?

Extended Writing

4 Kasia is a vegetarian. She says if everyone ate cereals instead of meat we could feed more people. Explain why she is right.

A02

5 The genes that code for an important human protein have been identified. Now scientists want to genetically engineer goats to produce it in their milk. List the steps the scientists need to take to make large quantities of the protein.

A01

6 Lamarck and Darwin both proposed theories that explained evolution. Using giraffes as an example, explain the similarities and differences between their theories.

A03

A01	Recall the science
A02	Apply your knowledge
A03	Evaluate and analyse the evidence

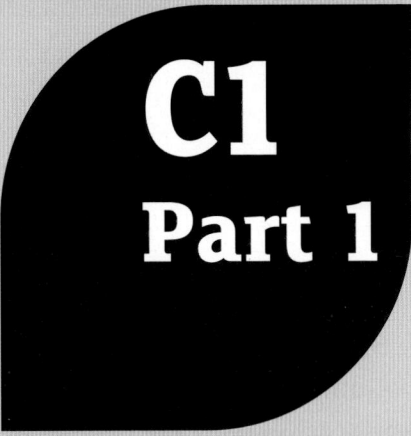

C1 Part 1

Atoms, rocks, metals, and fuels

Why study this unit?

The Earth's crust provides many resources for us. Extracting metals, rocks, and oil products can damage the environment and create dangerous waste.

In this unit you will discover how to extract metals from the Earth's crust, and how to make metal alloys with perfect properties for particular purposes. You will learn where limestone comes from, and about the chemical reactions that produce cement and concrete from this vital raw material. You will find out about fossil fuels too, and discover how the compounds of crude oil are separated to make petrol, diesel, and jet fuel.

You should remember

1 Everything is made up of tiny particles called atoms.

2 In chemical reactions, atoms are rearranged. Energy is given out or taken in as new products are made.

3 There are patterns in the chemical reactions of substances such as metals.

4 The way we use materials depends on their properties.

5 There are patterns in the properties of materials such as metals.

6 Humans extract many useful materials, such as metals and oil, from the Earth's crust.

The 92-storey Trump International Hotel and Tower in Chicago is the tallest residential building in the United States. The tower was made from about 330 000 tonnes of concrete, 22 500 tonnes of steel, and many tonnes of glass. During construction, a huge pump forced liquid concrete to the upper layers of the tower at a rate of over 2700 kg per minute.

Learning objectives

After studying this topic, you should be able to:

✔ know what elements are and what they are made up of

✔ describe how elements are classified in the periodic table

Key words

metal, element, atom, symbol, periodic table, group

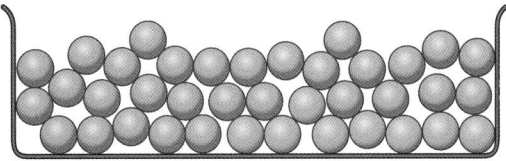

▲ A model of the atoms in liquid mercury. Each circle represents one mercury atom.

Did you know...?

You are made up of atoms of elements, too. A 50 kg person contains 32.5 kg of oxygen atoms, 9 kg of carbon atoms, 5 kg of hydrogen atoms, and smaller amounts of many other elements.

A What is an element?

B What is an atom?

Killer cargo

It's 1810. A ship takes on board a cargo of mercury. Within three weeks, many crew members are ill. Most are dribbling uncontrollably. Many have mouth ulcers and bowel complaints. Some are suffering more badly. Their faces are so swollen that their eyes will not open. Their tongues are so swollen that they can hardly breathe. Three sailors die. So do all the sheep, pigs, and cats on the boat.

These sailors and animals were suffering from mercury poisoning. The leather of the mercury containers had rotted. Liquid mercury flowed all over the ship. Sailors breathed in mercury vapour and absorbed the **metal** through their skin.

▲ Mercury is the only metal which is liquid at room temperature

Using mercury

Mercury is not all bad. Its vapour is a vital part of low energy light bulbs – each bulb contains about five milligrams of the metal. Mercury saves lives in tip-over switches in electric heaters, too. When the heater is on, electricity flows through the mercury in the switch. If the heater falls over, the mercury moves. The circuit is broken, so the heater switches off and is unlikely to start a fire.

Elements and atoms

Mercury is an **element**. It is made of just one sort of atom. An **atom** is the smallest part of an element that can exist. Atoms are tiny. The diameter of one atom is about 0.000 000 01 cm. If you could place one hundred million atoms side by side, they would stretch one centimetre.

Mercury is not the only element. In total, there are about 100 elements, each with its own type of atom. You cannot split elements into simpler substances. Everything in the world is made from the atoms of one or more of these 100 or so elements.

Symbols for elements

Each element has its own **symbol**. The symbol for mercury is Hg. This comes from its Latin name, *hydrargyrum*. The symbol O represents one atom of oxygen. The list below shows the symbols of some other elements:

- nitrogen – N
- neon – Ne
- neptunium – Np
- sodium – Na.

C Name an element whose symbol is the first letter of its name.

D Suggest a reason for the symbols of neon and neptunium being two letters long.

The periodic table

All the elements are listed in the **periodic table**. The vertical columns are called **groups**. The elements in a group have similar properties.

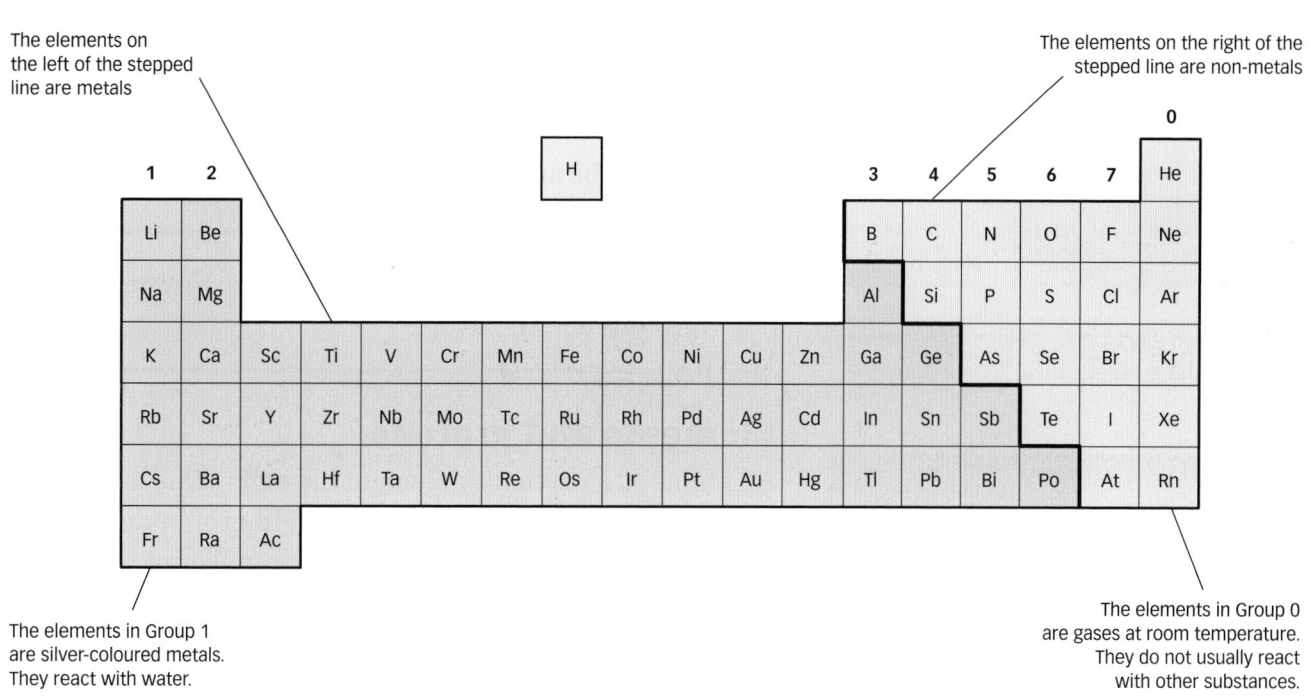

The elements on the left of the stepped line are metals

The elements on the right of the stepped line are non-metals

The elements in Group 1 are silver-coloured metals. They react with water.

The elements in Group 0 are gases at room temperature. They do not usually react with other substances.

Questions

1 Give the name and symbol of one element in Group 1 of the periodic table.

2 State what all the elements in Group 0 have in common.

3 How many types of atom are there? Explain how you worked out your answer.

Exam tip   AQA

✔ Remember that everything is made up of atoms of about 100 elements.

Learning objectives

After studying this topic, you should be able to:

✔ describe what atoms are made up of

✔ describe how electrons are arranged in atoms

✔ link electron arrangements to periodic table groups

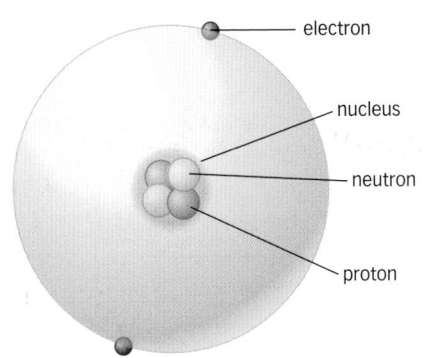

▲ Protons and electrons are electrically charged. Neutrons have no electrical charge. They are neutral.

Name of particle	Charge
proton	+1
neutron	0
electron	−1

Electronic structure ▶ of hydrogen

Electronic structure ▶ of lithium

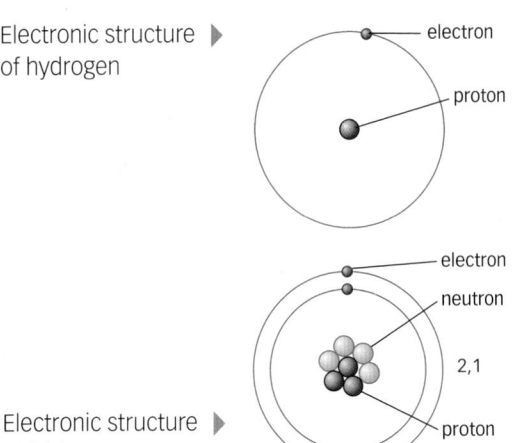

Exploring atoms

For many years, scientists believed that atoms were solid particles, like miniature snooker balls. But by the late 1800s, some scientists were beginning to doubt this idea. Scientists such as J. J. Thomson and Ernest Rutherford wanted to work out what atoms are really like. They carried out experiments to gather evidence.

Inside atoms

We now know that an atom is mainly empty space. At its centre is a **nucleus**. The nucleus is made up of tiny particles called **protons** and **neutrons**. Outside the nucleus are even tinier particles, called **electrons**.

Overall, an atom has no electrical charge. This is because it has equal numbers of positive protons and negative electrons. An oxygen atom, for example, has eight protons and eight electrons. So the atom has no overall electrical charge.

> **A** A nitrogen atom has seven electrons. How many protons does it have?

Elements and protons

All atoms of a particular element have the same number of protons. So every oxygen atom has eight protons, and every nitrogen atom has seven protons. The number of protons in an atom of an element is its **atomic number**. Atoms of different elements have different numbers of protons. The sum of the protons and neutrons in an atom is its **mass number**.

Arranging electrons

Electrons are arranged in **energy levels**. Each electron in an atom is in a particular energy level. Electrons fill the lowest energy levels first.

Hydrogen has just one electron. It occupies the lowest available energy level.

Lithium has three electrons. Two electrons fill up the lowest energy level. The other electron goes in the next energy level.

The electrons in sodium and potassium are arranged like this.

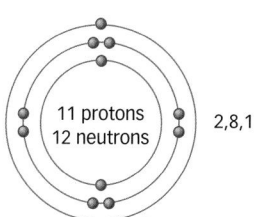

▲ Electronic structure of sodium

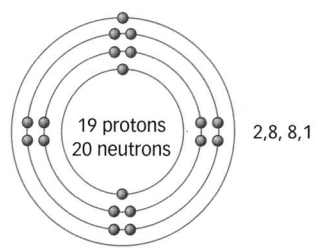

▲ Electronic structure of potassium

In all atoms, the lowest energy level can hold a maximum of two electrons. The next energy level holds up to eight electrons.

> B Neon has 10 electrons and argon has 18. Draw the electrons in an atom of neon and in an atom of argon.

Electron arrangements and the periodic table

The elements lithium, sodium, and potassium are in Group 1 of the periodic table. The atoms of each of these elements have just one electron in their highest energy level.

Similar electron arrangements give Group 1 elements similar properties. Each reacts vigorously with water to make hydrogen gas and a metal hydroxide. For example

$$potassium + water \rightarrow potassium\ hydroxide + hydrogen$$

◀ Potassium reacts vigorously with water

Group 1 metals also have similar reactions when burned in air. They all react vigorously with oxygen to make a metal oxide.

$$sodium + oxygen \rightarrow sodium\ oxide$$

The elements neon, argon, and krypton are **noble gases**. They are in Group 0 of the periodic table. Their atoms have eight electrons in the highest energy level. Helium is in Group 0 too. Its atoms have two electrons in its highest energy level. These electron arrangements are very stable. So the noble gases are **unreactive**.

Key words

nucleus, proton, neutron, electron, atomic number, mass number, energy level, noble gas, unreactive

Did you know...?

If you could make an atom as big as a football stadium, its nucleus would be as small as a grain of sand.

Exam tip   AQA

✔ It might help to think of electron energy levels as shells. The lowest energy level is the innermost shell.

Questions

1 An atom of neon has ten protons. How many electrons does it have?   ↓ E

2 Draw diagrams to show the electron arrangements of carbon (six electrons), oxygen (eight electrons), and neon (ten electrons).   ↓ C

3 An atom of argon has 18 protons and 22 neutrons. Work out its atomic number and its mass number.

4 Use ideas about electron arrangement to explain why the elements in Group 0 have similar properties.   ↓ A*

Learning objectives

After studying this topic, you should be able to:

- ✔ explain the differences between a compound and an element
- ✔ explain ionic and covalent bonding

▲ If you eat too much salt you are more likely to get heart disease

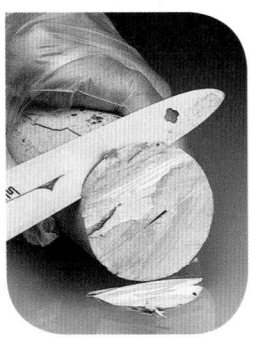

▲ Sodium

▲ Chlorine

▲ Sodium chloride

Vital element

Sodium is an element. It is vital to life. You need sodium to make your heart beat properly and to help your nerves transmit messages.

Of course, you can't eat pure sodium – the metal reacts violently with water. Instead you need to eat substances that contain sodium, such as sodium chloride, or salt.

Compounds

Sodium chloride is a **compound.** Compounds are made up of two or more elements, strongly joined together.

The properties of compounds are different from the properties of the elements from which they are made. For example:

- Sodium is a solid silver-coloured metal at room temperature. It conducts electricity.
- Chlorine is a green gas. It is a non-metal. It does not conduct electricity.
- Sodium chloride is a white solid. It does not conduct electricity when solid.

A What is a compound?

B Describe two ways in which the compound sodium chloride is different from each of the elements sodium and chlorine.

Holding compounds together
Ionic bonds

Sodium chloride is made up of a metal joined to a non-metal. When sodium chloride forms from its elements, each sodium atom transfers one of its electrons to a chlorine atom.

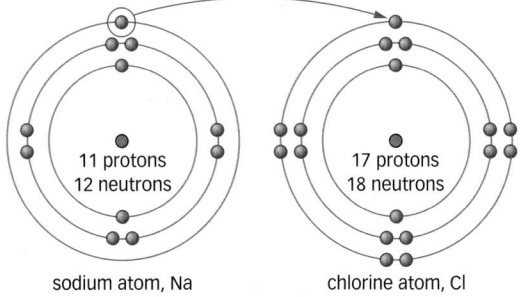

11 protons
12 neutrons

17 protons
18 neutrons

sodium atom, Na   chlorine atom, Cl

Each atom now has eight electrons in its highest energy level. These electron arrangements are very stable.

- The sodium atom now has 10 electrons for its 11 protons. Overall, it has a charge of +1.
- The chlorine atom now has 18 electrons and 17 protons. Overall, it has a charge of –1.

Charged atoms are called **ions**. Here, we have formed a positive sodium ion and a negative chloride ion.

Sodium chloride is an **ionic compound**. In ionic compounds, the positive and negative ions are strongly attracted to each other. The forces of attraction between positive and negative ions are called **ionic bonds**. A sodium chloride crystal consists of millions of sodium and chloride ions, all held together in a regular pattern by ionic bonds.

Ionic bonds only exist in compounds made up of metals and non-metals.

Covalent bonds

The compound carbon dioxide is made up of two non-metals. So it cannot contain ionic bonds – there are no metal atoms to give away electrons.

Instead, its atoms are joined together in groups of three. Each group has one carbon atom and two oxygen atoms. The atoms share electrons between them to form **covalent bonds**.

Other compounds made up of non-metals only are held together by covalent bonds. The atoms in gases such as oxygen, O_2, and chlorine, Cl_2, also share electrons to form covalent bonds.

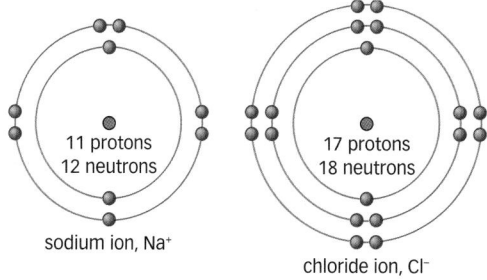

11 protons
12 neutrons

sodium ion, Na$^+$

17 protons
18 neutrons

chloride ion, Cl$^-$

C What is an ionic bond?

D Potassium chloride is held together by ionic bonds. Predict the charges on the potassium and chloride ions. Explain your predictions.

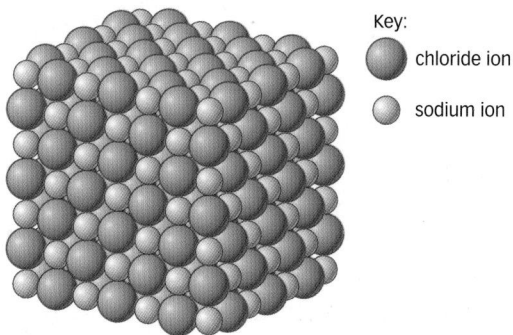

Key:
⬤ chloride ion
◯ sodium ion

▲ Sodium ions and chloride ions are arranged like this

Key words

compound, ion, ionic compound, ionic bond, covalent bond

Exam tip   AQA

✔ Remember – compounds made up of a metal and a non-metal are held together by ionic bonds. Compounds made up of non-metals only are held together by covalent bonds.

Questions

1 How many elements join together to make the compound sodium chloride?

2 Use ideas about atoms to explain the difference between an element and a compound.

3 Explain how sodium and chloride ions are formed when sodium chloride is made from its elements.

4 Explain how the atoms in a carbon monoxide molecule are held together.

Learning objectives

After studying this topic, you should be able to:

- ✔ understand what happens in chemical reactions
- ✔ write word equations to summarise reactions
- ✔ interpret and write chemical formulae

Key words

chemical reaction, reactant, product, formula

A Describe two features of chemical reactions.

B Carbon burns in oxygen to make carbon dioxide. Write a word equation for the reaction.

C Methane burns in oxygen to make carbon dioxide and water. Write a word equation for the reaction.

D Copper carbonate decomposes on heating to make copper oxide and carbon dioxide. Write a word equation for the reaction.

Chemical reactions

A teacher heats a small piece of sodium. When the sodium catches fire, she places it in a flask of chlorine. The sodium continues to burn. White fumes are produced.

◀ Sodium reacts vigorously with chlorine

There has been a **chemical reaction**. Atoms of sodium and chlorine have joined together to make a new substance, sodium chloride.

There are millions of possible chemical reactions. In all of them, the atoms of the starting materials, or **reactants**, are rearranged to make new substances, or **products**. The properties of the products are different from those of the reactants.

Most chemical reactions are irreversible. Once a reaction has happened, it is difficult to get the starting materials back again.

Word equations

The teacher uses a word equation to summarise the reaction of sodium and chlorine:

sodium + chlorine → sodium chloride

Word equations show the reactants and products of chemical reactions. But they do not tell us much else. To explain how the atoms are rearranged in a reaction, or to work out the amounts of substances that react together, you need a symbol equation.

Formulae

Before you can write a symbol equation, you need to know the symbols or formulae of the reactants and products. For the reaction of sodium with chlorine:

- The symbol of the element sodium is Na.
- The symbol of the element chlorine is Cl. In chlorine gas the atoms are joined together in pairs to make chlorine molecules. So we represent chlorine gas by the formula Cl_2.
- The formula of sodium chloride is NaCl. This shows that the compound is made up of atoms of two elements. For every atom of sodium, there is one atom of chlorine. The **formula** tells us the number and type of atoms that are joined together in the compound.

Each compound has its own formula. The formula of carbon dioxide is CO_2. This shows that a molecule of the compound is made up of atoms of two elements, carbon and oxygen. For every one atom of carbon, there are two atoms of oxygen.

▲ A molecule of chlorine consists of two chlorine atoms strongly joined together

Key:
- oxygen atom
- carbon atom

▲ A molecule of carbon dioxide consists of one carbon atom and two oxygen atoms

---

E   A molecule of nitrogen gas consists of two nitrogen atoms joined together. Write the formula of nitrogen gas.

F   The compound lithium fluoride is made up of one lithium ion for every fluoride ion. Write the formula of lithium fluoride.

Questions

1   Write a word equation for the reaction of potassium with fluorine to make potassium fluoride.

2   Write the symbol of the element chlorine.

3   Write the formula of chlorine gas.

4   A molecule of sulfur trioxide consists of one atom of sulfur joined to three atoms of oxygen. Work out its formula.

5   The formula of potassium manganate(VII) is $KMnO_4$. Explain what the formula tells us about this compound.

↓ E

↓ C

↓ A*

Exam tip   AQA

- ✓ You cannot guess the formulae of compounds. Look them up in a book or on the Internet. Learn the formulae of common compounds such as sodium chloride and carbon dioxide.

Learning objectives

After studying this topic, you should be able to:

✔ appreciate that mass is conserved in chemical reactions

✔ interpret symbol equations

✔ write balanced symbol equations

Key words

balanced symbol equation

▲ Burning titanium has started several fires in military jet plane engines

Titanium fire

Titanium metal is important in the aerospace industry. Most of its properties make it an ideal material for jet engine parts. There's just one problem. At high temperatures, it catches fire. And moving parts in jet engines quickly reach high temperatures.

Symbol equations

The word equation for the burning reaction of titanium is

titanium + oxygen → titanium dioxide

Atoms cannot be lost or made in chemical reactions. So if 48 g of titanium reacts with 32 g of oxygen, the mass of titanium dioxide formed will be 48 g + 32 g = 80 g.

A **balanced symbol equation** tells us more about the reaction by showing:

• how the atoms are rearranged
• the relative amounts of the substances that take part in the reaction.

Here's how to write a balanced symbol equation for the reaction of burning titanium:

• Write a word equation for the reaction. Put the correct symbol or formula under each reactant and product.

titanium + oxygen → titanium dioxide

$Ti + O_2 \rightarrow TiO_2$

• Now balance the equation. The equation must show the same amounts of each substance on both sides of the arrow. Here, there are two atoms of oxygen and one atom of titanium on each side of the arrow. The equation is balanced.

A List two things that a balanced symbol equation shows.

B The balanced symbol equation for the burning reaction of carbon to make carbon dioxide is $C + O_2 \rightarrow CO_2$. Explain what the equation tells us about the reaction.

Laptop fire

It's 2006. Marv is using his laptop computer. Suddenly, it catches fire. Marv quickly extinguishes the flames, but the computer is ruined. What caused the fire?

It turns out that Marv wasn't alone. Many laptops caught fire at around the same time. Engineers realised that their lithium batteries were to blame. Sometimes, the batteries short-circuited, causing a spark. The spark set fire to tiny pieces of lithium metal floating in the battery's liquid. Then the fire spread.

Here's how to write a balanced symbol equation for the reaction:

- Write a word equation. Write a symbol or formula under each substance.

$$\text{lithium} + \text{oxygen} \rightarrow \text{lithium oxide}$$
$$Li + O_2 \rightarrow Li_2O$$

- Balance the amounts of oxygen. There are two atoms of oxygen on the left of the arrow and one on the right. Write a big number 2 to the left of the formula of lithium oxide.

$$Li + O_2 \rightarrow 2Li_2O$$

- The big 2 applies to each type of atom in the formula that follows it. Here, it means there are $(2 \times 2) = 4$ atoms of lithium and $(1 \times 2) = 2$ atoms of oxygen. The numbers of oxygen atoms are now balanced.
- Balance the amounts of lithium by writing a big 4 to the left of the symbol for lithium. There are now 4 atoms of lithium on each side. The equation is balanced.

$$4Li + O_2 \rightarrow 2Li_2O$$

Exam tip AQA

- ✔ Never alter a formula to balance an equation. Add big numbers to the left of the formulae as needed.

Questions

1 Write a word equation for the burning reaction of sulfur to make sulfur dioxide, SO_2. ↓ E

2 Describe in words what the equation below tells us about the reaction of magnesium with oxygen. ↓ C
$$2Mg + O_2 \rightarrow 2MgO$$

3 Sodium burns in oxygen to make sodium oxide, Na_2O. Write a balanced symbol equation for the reaction.

4 Write a balanced symbol equation to show that magnesium reacts with hydrochloric acid, HCl, to make hydrogen gas, H_2, and magnesium chloride, $MgCl_2$. ↓ A*

Learning objectives

After studying this topic, you should be able to:

✔ explain how the properties of limestone make it a good building material

✔ evaluate the environmental, social, and economic effects of exploiting limestone

▲ Limestone buildings in Royal Crescent, Bath

▲ This limestone statue has been damaged by acid rain

Rock of beauty

Every year, more than a quarter of a million people visit the city of Bath in southwest England. They are drawn to its Roman baths and its beautiful honey-coloured buildings. Many of these buildings were constructed more than 200 years ago. Most residents welcome the tourists – they bring in money and sustain jobs.

The houses and shops of the city are made from Bath stone. Bath stone is a type of **limestone**. Limestone was formed from the shells of creatures that lived in shallow seas between 440 and 70 million years ago.

Limestone is an important building material. It is attractive, durable, and strong. It can be cut into blocks. Some limestone is crushed into small lumps to make **aggregate**. Aggregate is used as a firm base beneath railway lines and roads.

Limestone is also a raw material for the production of other useful building materials, including cement, mortar, and glass.

Acid attack

Unfortunately the surface of limestone can be damaged by chemical reactions with acid rain (see spread C1.15). This may make gaps between blocks in buildings.

The damage happens because limestone consists mainly of calcium carbonate, $CaCO_3$.

Calcium carbonate reacts with acids to make a salt, water and carbon dioxide. For example with sulfuric acid, an acid rain acid, the equation is:

$$\text{calcium carbonate} + \text{sulfuric acid} \rightarrow \text{calcium sulfate} + \text{carbon dioxide} + \text{water}$$

$$CaCO_3 + H_2SO_4 \rightarrow CaSO_4 + CO_2 + H_2O$$

Other carbonates react in similar ways with acids. For example:

$$\text{magnesium carbonate} + \text{hydrochloric acid} \rightarrow \text{magnesium chloride} + \text{carbon dioxide} + \text{water}$$

A Describe three properties of limestone. Explain why these properties make limestone a good building material.

Quarry queries

Companies get limestone from quarries – large holes in the ground. They use explosives to break up the rock before digging it out. Quarries bring benefits and problems.

Benefits

Many quarries are in the countryside. Here, they provide jobs in places where work may be scarce. This helps local families and facilities such as shops and schools. These are **social impacts** of quarrying.

Products from quarries are valuable. Each year, British quarries produce materials worth several billion pounds. Although the UK imports some limestone, it exports even more. This contributes to the nation's economy. These are **economic impacts** of digging rock from the ground.

Problems

Some quarries are in attractive areas of the countryside, where they may damage the tourist industry. Transporting rock from quarries to customers creates extra traffic, which may pass through small towns or villages.

Quarries also have **environmental impacts**. For example they take up land space, making the land unavailable for other uses such as farming and recreation.

> B Describe one social benefit and one social problem caused by quarrying.
>
> C Describe an economic impact of quarrying limestone.

▲ Limestone quarry

Did you know...?

In the UK, we produce about 80 million tonnes of the rock every year. That's more than one tonne for every man, woman, and child.

Key words

limestone, aggregate, social impact, economic impact, environmental impact

Exam tip   AQA

✓ Always relate uses of materials to their properties.

Questions

1 Calcium carbonate is the main mineral in limestone. Write down its formula and explain what it means. ↓E

2 Write a word equation to summarise the reaction of calcium carbonate. ↓C

3 Describe the social, economic, and environmental impacts of quarrying limestone. Explain which impacts are benefits and which are problems. ↓A*

Learning objectives

After studying this topic, you should be able to:

✔ describe how to make calcium oxide, calcium hydroxide, and limewater and explain why they are useful

✔ describe and explain the lime cycle

▲ Heating limestone produces solid calcium oxide and carbon dioxide gas

Key words

thermal decomposition, state symbols, limewater, lime cycle

Did you know...?

Calcium hydroxide (lime) is an ingredient of hair relaxers and some hair removal creams.

Heated calcium oxide glows so brightly it used to be used for stage lighting. Leading actors were 'in the limelight'.

Making calcium oxide

Nadeem weighs a lump of limestone. He heats it for five minutes in a Bunsen flame. The limestone gets so hot that it glows. Nadeem lets the rock cool down. Then he weighs it again. The mass has decreased. Why?

On heating, calcium carbonate breaks down to make two new materials – calcium oxide and carbon dioxide gas. This is a **thermal decomposition** reaction.

calcium carbonate → calcium oxide + carbon dioxide

The balanced symbol equation for the reaction is:

$$CaCO_3(s) \rightarrow CaO(s) + CO_2(g)$$

The (s) shows that calcium carbonate and calcium oxide are both solids under the conditions of the reaction. Carbon dioxide is formed as a gas. The (s) and (g) are **state symbols**.

> A Name the products of the thermal decomposition reaction of calcium carbonate.
>
> B Give the states of each of the products of the reaction.

Making calcium hydroxide

Nadeem goes back to his calcium oxide. He adds water to it, drop by drop. There is a chemical reaction. The product is calcium hydroxide. The reaction gives out so much heat it makes the water boil as Nadeem adds it.

calcium oxide + water → calcium hydroxide

$$CaO(s) + H_2O(l) \rightarrow Ca(OH)_2(s)$$

The (l) shows that the water is liquid.

> C Give the meanings of the symbols (s), (l), and (g) in symbol equations.

Completing the lime cycle

Nadeem adds more water to the calcium hydroxide. Some of the calcium hydroxide dissolves. Nadeem filters the mixture and collects a colourless solution of calcium hydroxide. The solution is also called **limewater**.

Limewater is the test for carbon dioxide gas. Nadeem blows into the limewater through a straw. The limewater goes cloudy. The tiny pieces of solid that make the solution look cloudy are calcium carbonate. Nadeem has made the material he started with.

calcium hydroxide	+	carbon dioxide	→	calcium carbonate	+	water
$Ca(OH)_2(aq)$	+	$CO_2(g)$	→	$CaCO_3(s)$	+	$H_2O(l)$

The (aq) shows that the calcium hydroxide is dissolved in water.

Nadeem's series of reactions together make the **lime cycle**. The diagram below summarises this cycle.

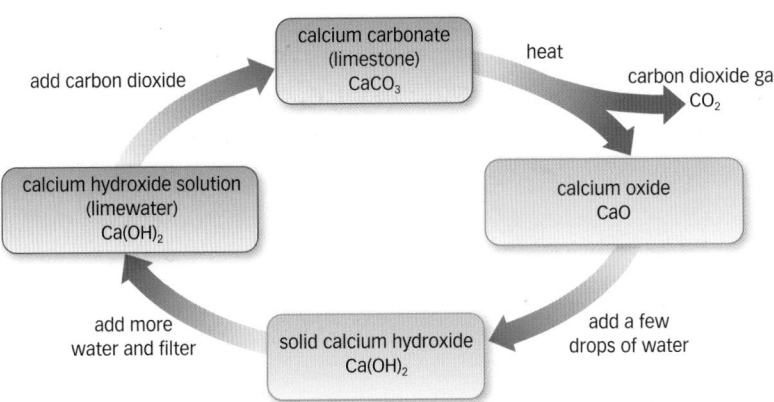

Using products from the lime cycle

Calcium oxide and calcium hydroxide both form alkaline solutions with water. They have many uses:

• neutralising excess acidity in lakes and soils
• neutralising acidic waste gases produced by burning coal in power stations (see spread C1.15).

◄ Calcium oxide and calcium hydroxide neutralise acidic lakes

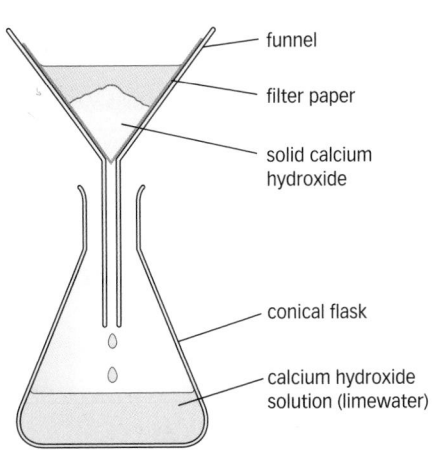

▲ Filtering calcium hydroxide

Exam tip **AQA**

✓ Learn the reactions of the lime cycle.

Questions

1 Give the formulae of calcium oxide and calcium hydroxide.

2 Name the type of reaction that happens if you heat calcium carbonate strongly.

3 Copy the diagram of the lime cycle. Add diagrams to show what Nadeem did at each stage to make the reactions happen.

4 Give the meaning of the symbol (aq) in symbol equations.

5 Calculate the mass of carbon dioxide produced in a thermal decomposition reaction if you start with 10.0 g of calcium carbonate and make 5.6 g of calcium oxide.

Learning objectives

After studying this topic, you should be able to:

✔ describe how cement, mortar, and concrete are made from limestone

✔ describe and explain the pattern in the decomposition reactions of metal carbonates

Key words

cement, mortar, concrete, reinforced concrete

Did you know...?

Cement-making accounts for about 5% of the carbon dioxide humans send into the atmosphere.

A List the raw materials for making cement.

B Suggest one reason for heating the kiln when making cement.

Introduction

What do these structures have in common: the world's tallest building (the Burj Khalifa tower in Dubai), a brick house in the UK, and water pipes?

▲ Cement – a dry grey powder – was used in the construction of them all

Cement

Companies make **cement** like this:
- Crush limestone rock into small pieces.
- Add powdered clay.
- Heat the mixture to 1450 °C in a rotating kiln.
- Add a little calcium sulfate powder.

During the heating stage, calcium carbonate in the limestone decomposes to make calcium oxide (quicklime) and carbon dioxide.

Mortar

Bob is a bricklayer. He uses **mortar** to stick bricks together. He makes the mortar by mixing a thick paste from cement, sand, and water. Mortar sets overnight as substances in the mixture react with each other.

◀ Bricklayers use mortar to stick bricks together

Concrete

Builders used 110 000 tonnes of **concrete** to make the foundations of the Burj Khalifa tower. They used an extra 39 000 tonnes of concrete to build the rest of the tower. The foundations and the tower are reinforced by huge steel bars.

Construction workers make concrete by mixing cement, sand, aggregate (small stones), and water. Like mortar, concrete sets as a result of chemical reactions within the mixture.

Concrete can form structures of many different shapes. It is strong under forces of compression (squashing), but weak if bent or stretched. It can be made much stronger by reinforcing it with steel. Because of these properties, concrete is the main material in millions of buildings. Bridges, mains water pipes, and paths are also made from concrete or **reinforced concrete**.

Decomposing metal carbonates

Calcium carbonate decomposes when limestone is heated to make cement. The carbonates of sodium, magnesium, zinc, and copper also break down in thermal decomposition reactions.

For example, copper carbonate produces two products when you heat it strongly – copper oxide and carbon dioxide. Copper carbonate is green and copper oxide is black.

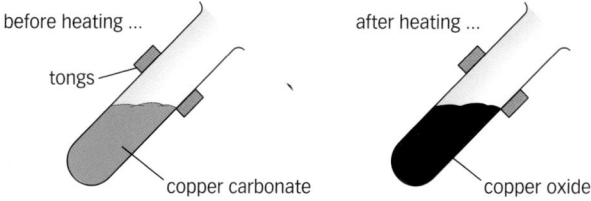

before heating ...   tongs   copper carbonate

after heating ...   copper oxide

The equation for the reaction is:

$$\text{copper carbonate} \rightarrow \text{copper oxide} + \text{carbon dioxide}$$
$$\text{CuCO}_3(s) \rightarrow \text{CuO}(s) + \text{CO}_2(g)$$

Thermal decomposition reactions follow a pattern.

- Carbonates of metals low in the reactivity series (see spread C1.9), such as copper carbonate, need relatively small amounts of energy to break them down.
- The carbonates of metals high in the reactivity series, such as sodium carbonate, need a lot of energy to make them decompose.
- The carbonates of very reactive metals, such as potassium carbonate, do not decompose at Bunsen burner temperatures.

C Describe two advantages of using concrete as a building material.

D Explain why concrete structures have steel bars inside them.

Exam tip

✔ Remember the raw materials for making cement, mortar, and concrete.

E Describe one change you would see if you heated copper carbonate strongly.

Questions

1 Describe the appearance of cement.

2 List the raw materials used for making mortar and concrete.

3 What happens inside mortar and concrete to make them harden?

4 Write a word equation for the thermal decomposition reaction of sodium carbonate.

5 Describe the pattern shown by the thermal decomposition reactions of metal carbonates.

6 Write a balanced symbol equation for the thermal decomposition reaction of zinc carbonate, ZnCO_3.

Learning objectives

After studying this topic, you should be able to:

✔ describe typical transition metal properties

✔ describe and explain how to extract gold and iron from the Earth's crust

Metals for building

Imagine watching a football match in an aluminium-clad stadium, or enjoying art in Spain's titanium-covered museum. Or how about learning science in a copper-clad building in the Arctic? The properties of the metals used make them ideal for these buildings.

> **A** Suggest why the architects chose to cover the buildings in these metals.

▲ The copper-clad science centre in Svalbard, Norway

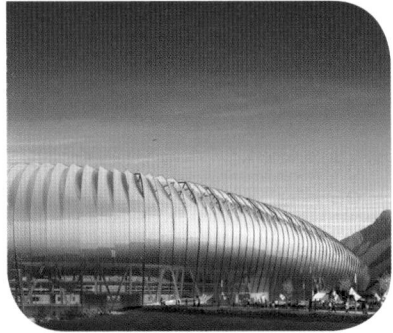

▲ The aluminium-clad Monterrey stadium in Mexico

▲ Spain's titanium-covered Guggenheim museum

Did you know...?

Most gold jewellery is not pure gold. It is mixed with other metals to make it harder.

Key words

transition metal, unreactive, reactivity series, mineral, ore, blast furnace, reduction

Metals in the middle

Titanium and copper are in the central block of the periodic table. Together, all the metals in this block are the **transition metals**. Like most metals, transition metals:

- have a shiny surface when freshly cut
- can be bent or hammered into different shapes without cracking
- are good conductors of heat and electricity.

1	2											3	4	5	6	7	0
						H											He
Li	Be											B	C	N	O	F	Ne
Na	Mg											Al	Si	P	S	Cl	Ar
K	Ca	Sc	Ti	V	Cr	Mn	Fe	Co	Ni	Cu	Zn	Ga	Ge	As	Se	Br	Kr
Rb	Sr	Y	Zr	Nb	Mo	Tc	Ru	Rh	Pd	Ag	Cd	In	Sn	Sb	Te	I	Xe
Cs	Ba	La	Hf	Ta	W	Re	Os	Ir	Pt	Au	Hg	Tl	Pb	Bi	Po	At	Rn
Fr	Ra	Ac															

transition metals

Where do metals come from?

Gold

Why was gold one of the first metals to be discovered and used? Probably because it was easy to find – humans couldn't miss the grains and nuggets of the metal shining in the stream beds of ancient Egypt and the area that is now Iraq.

Gold is found as the metal itself because it is **unreactive**. It hardly ever joins to other elements to form compounds. It is near the bottom of the **reactivity series**.

Iron

Iron – like most metals – is too reactive to exist on its own in the Earth's crust. It is joined to other elements in naturally occurring compounds called **minerals**. One important iron mineral is haematite – iron(III) oxide, Fe_2O_3. Minerals do not usually exist alone. They are mixed with sand or rock. Rocks that contain useful minerals are **ores**.

Iron is an important metal. Every year companies dig millions of tonnes of iron ore from the ground. They use chemical reactions to extract iron metal from the ore. Here's how:

- Put the iron ore (mainly iron(III) oxide) in a hot **blast furnace** with coke (carbon).
- Oxygen is removed from the iron(III) oxide in **reduction** reactions. The products are iron and carbon dioxide.

Other metals

Other metals below carbon in the reactivity series are also extracted by heating their oxides with carbon. Carbon reduces metal oxides. For example:

tin oxide + carbon → tin + carbon dioxide

most reactive

sodium
calcium
magnesium
aluminium
carbon
zinc
iron
tin
lead
copper
silver
gold
platinum

least reactive

▲ The reactivity series lists metals in order of their reactivity. Carbon is included to show its reactivity, even though it is not a metal.

Exam tip **AQA**

✓ Metals that are less reactive than carbon can be extracted from their oxides by reduction with carbon.

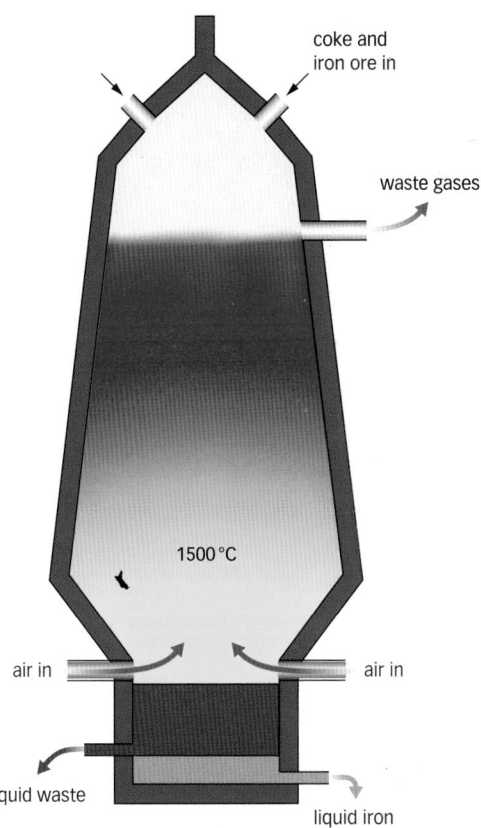

coke and iron ore in

waste gases

1500 °C

air in   air in

liquid waste

liquid iron

▲ Iron oxide is reduced in the blast furnace to make iron

Questions

1 List three typical properties of transition metals.
2 Name two metals that are extracted from their minerals by heating with carbon.
3 What is a reduction reaction?
4 Write a word equation for the reduction of lead oxide by carbon to make lead and carbon dioxide.

↓ E

▼ ◄ ►

C

↓ A*

Learning objectives

After studying this topic, you should be able to:

✔ describe how atom arrangements in iron and steels are linked to their properties and uses

✔ explain the meaning of the word alloy

▲ A cast iron cannon

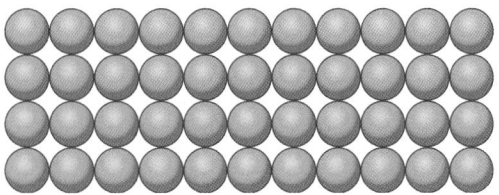

▲ The diagram shows one way of modelling the atom arrangement in pure iron

Key words

cast iron, steel, alloy, low carbon steel, high carbon steel, corrosion, stainless steel

A Explain why pure iron is not useful.

B Explain why steel is an example of an alloy.

Issues with iron

Iron from the blast furnace is not pure. It is a mixture of about 96% iron, 3% carbon, and other impurities. The impurities make blast furnace iron brittle – it breaks easily when you drop it. Blast furnace iron is not much use as it is.

Cast iron

Re-melting blast furnace iron and adding scrap steel makes **cast iron**. Cast iron has a high strength in compression – you can press down on it with a great force and it will not break. Because of this property, cast iron was made into cannons and cooking pots.

Pure iron

Removing the impurities from blast furnace iron makes pure iron. Pure iron has a regular arrangement of atoms.

The layers of atoms in pure iron slide over each other easily. This makes it easy to bend it into different shapes. Pure iron is also soft, which means it is not very useful.

So why do companies extract iron from its ores? How do they change its properties to make it useful?

Steel – a vital alloy

Steel makes many things – from stunning structures to tiny components.

▲ The Millau viaduct is made from steel

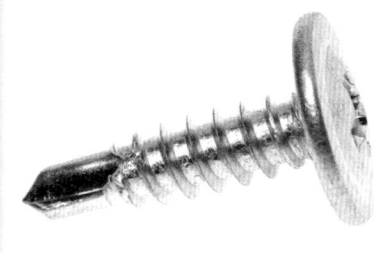

▲ A steel screw

Steel is mainly iron. The iron is mixed with certain amounts of carbon, and sometimes other metals, to change its properties and make it more useful. There are many types of steel.

Steels are examples of **alloys**. An alloy is a mixture of a metal with one or more other elements. The physical properties of an alloy are different from the properties of the elements in it.

Inside steel

The properties of an alloy depend on its atom arrangement. The atoms of carbon, iron, and other metals are of different sizes. In steel, the carbon and other metal atoms get between the iron atoms. They distort the regular pattern. The layers of iron atoms can no longer slide over each other easily. So steel has different properties from pure iron. Steel is harder and less bendy.

Right for the job

People have been making steel for centuries. Bracelets that are 15 000 years old have been discovered in Tanzania. A thousand years ago, Chinese people made sharp steel swords.

Carbon steels

Over the years, scientists experimented with different mixtures to make steels with perfect properties for particular purposes. They found that **low carbon steels**, containing less than 0.3% carbon, are easy to make into different shapes. Steel companies make low carbon steel sheets for car body panels and food cans.

High carbon steels have different properties. They are hard and strong. High carbon steels contain between 0.6% and 1.0% carbon.

Stainless steel

Iron and steel go rusty when surface iron atoms react with water and oxygen from the air. The process is called **corrosion**. Corrosion weakens and damages iron and steel structures. It costs money and even lives.

Stainless steels do not go rusty. They are resistant to corrosion. They have this property because they contain chromium atoms.

Stainless steel makes cutlery, surgical instruments, and even the kitchen sink.

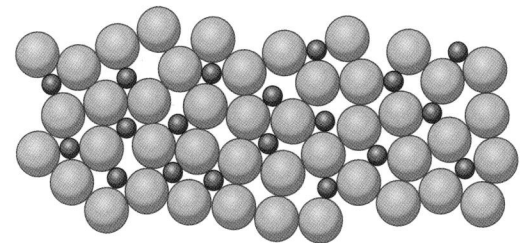
▲ A model of the structure of steel

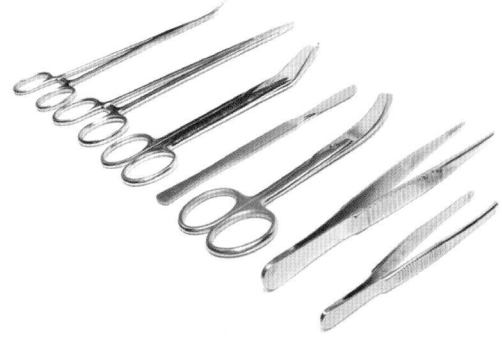
▲ These surgical instruments are made from stainless steel

Exam tip AQA

✓ Each type of steel has unique properties because of the arrangements of its atoms.

Questions

1 Use ideas about properties to explain why blast furnace iron is not useful.

2 Summarise the properties and uses of low carbon steel, high carbon steel, and stainless steel.

3 Use diagrams to explain how the atom arrangements of iron and steel explain their different properties.

Learning objectives

After studying this topic, you should be able to:

✔ explain how the properties of copper make it suitable for its uses

✔ describe how copper is extracted from its ores and from waste

Key words

open-cast mine, concentrate, smelting, electrolysis, low-grade ore, displacement, phytomining, bioleaching

A Give three uses of copper and explain how its properties make it suitable for each use.

An open-cast copper mine

Copper crooks

A gang dresses up in high visibility jackets. They set up a line of traffic cones and rip out underground cables. Criminals are stealing as much copper telephone cable as they can carry away. Why? What makes copper so valuable?

Pipes, pots, and pounds

Copper is valuable because its properties make it useful.

Chris is a plumber. He can choose either copper or plastic water piping – both are waterproof. Chris much prefers copper piping – he can bend it to the angles he needs and weld its joints to prevent leaks.

Sharon is an electrician. She uses many metres of plastic-covered copper wire every week, since copper is an excellent conductor of electricity.

Waqar is a cook. He uses copper-bottomed pans because copper conducts heat well.

Copper is also used to make coins. Five, ten, and twenty pence pieces are an alloy of copper with nickel. By itself, copper is too soft for coins. Mixing copper with nickel makes a harder alloy.

Copper from the Earth's crust

Most copper in the Earth's crust is joined to other elements in minerals. The minerals are mixed with other substances in ores. Companies use several processes as they extract copper from its ores:

- They dig the ore from the ground at an **open-cast mine**.
- They **concentrate** the ore to separate copper minerals from the waste rock of the ore.

The companies then obtain pure copper by either or both of these two processes:

- They heat the concentrated ore in a furnace. Chemical reactions remove other elements from the copper. This is **smelting**. They then purify the copper by **electrolysis**.
- They make a solution of copper compounds from the ore and obtaining copper metal by electrolysis.

Copper conundrum

Copper-rich ores are running out. So companies now extract copper from **low-grade ores** that contain little copper. This costs more than using copper-rich ores, but the high demand for copper means that they can still make a profit. But extracting copper from low-grade ores produces huge amounts of waste. The waste damages the environment.

These days, about half British copper is recycled, often by electrolysis. Copper can also be obtained from solutions of copper salts by adding scrap iron to the solutions. Chemical reactions result in the formation of copper metal. For example:

iron + copper sulfate → iron sulfate + copper

This is a **displacement** reaction.

Obtaining copper from solutions of its salts uses less energy and causes less pollution than extracting copper from its ores.

Scientists have also been researching new ways of extracting copper from low-grade ores without damaging the environment too much. The new techniques include:

- **Phytomining** – This involves planting certain plants on low-grade copper ores. The plants absorb copper compounds. Burning the plants produces ash that is rich in copper compounds.

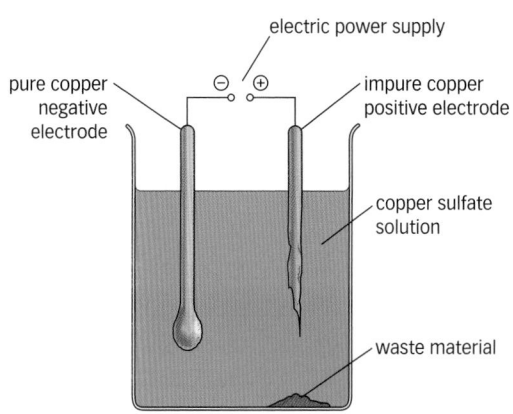

▲ The negative electrode is made of pure copper. Impure copper forms the positive electrode. During electrolysis, positive copper ions move to the negative electrode. Waste material falls to the bottom.

▲ The 'copper flower' plant absorbs large amounts of copper

- **Bioleaching** – Some bacteria can obtain their nutrients and energy from copper compounds in copper ores. The bacteria produce solutions of copper compounds. Chemical reactions or electrolysis extract copper metal from these solutions. The process is very slow.

Questions

1. Describe how copper is extracted from its ores.
2. Describe two problems connected to the extraction of copper from its ores.
3. Describe how copper is extracted by phytomining and bioleaching. Give an advantage of each process.
4. Suggest some economic impacts of the work of copper cable criminals.

12: Titanium and aluminium

Learning objectives

After studying this topic, you should be able to:

✔ give reasons for the uses of aluminium, titanium, and their alloys

✔ explain why the metals are expensive to extract

✔ explain the benefits of recycling metals

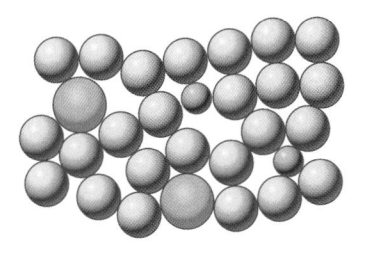

key:

 = iron atom   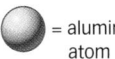 = aluminium atom    = silicon atom

▲ Atoms in an aluminium alloy

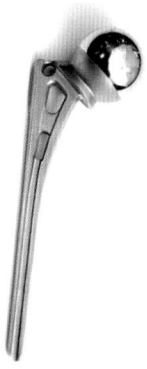

▲ An artificial hip made from titanium

> A Explain which properties of titanium make its alloys suitable for making aeroplane parts.
>
> B Use ideas about properties to suggest why aluminium is used to make overhead power cables.

Wheels and wings

Jason's car has alloy wheels. They were much more expensive than normal steel wheels. So why did he buy them?

First, Jason thinks shiny alloy wheels look good. He likes their exotic shapes and styles, too.

Second, the wheel alloy is mainly aluminium. Aluminium has a low density – it is light for its size. So alloy wheels are lighter than steel wheels of the same size.

Aluminium has a thin layer of aluminium oxide on its surface. This layer stops oxygen and water molecules reacting with the aluminium atoms underneath. So aluminium does not corrode.

Aluminium on its own would not make good wheels – the metal is too soft. But an alloy of 93% aluminium mixed with silicon and iron is perfect for the job. The silicon and iron atoms disrupt the pattern of atoms in pure aluminium. Aluminium atoms slide over each other less easily in the alloy, so the alloy is harder than the pure metal. Aluminium alloys are also much stronger than pure aluminium.

The properties of aluminium mean that the metal and its alloys are also used in:

- aeroplanes
- overhead power cables
- cooking foil
- drinks cans.

Using titanium

Titanium is a transition metal. It has typical transition metal properties. And, like aluminium, it has a low density and resists corrosion.

These properties mean that titanium alloys are useful for making aeroplanes, even though titanium catches fire more easily than some metals (see spread C1.5).

Titanium is also used to make artificial hips and bone pins, and for oil rigs at sea.

Extracting titanium and aluminium

Most of the aluminium we use came originally from aluminium oxide in bauxite ore. Most titanium exists naturally as titanium oxide. There are limited amounts of the ores of both metals – they won't last forever.

Aluminium and titanium are expensive. This is because they are not easy to extract. You can't extract aluminium or titanium from their ores by heating with carbon. Here's why:

- Aluminium is above carbon in the reactivity series. Its atoms are joined very strongly to oxygen atoms in bauxite.
- If you heat titanium oxide with carbon you make titanium carbide. This makes the metal brittle.

So aluminium is extracted by electrolysis. The process uses a lot of electrical energy, which is expensive.

Titanium is extracted from its ores in a multi-step process. Again, much energy is needed, so the process is expensive.

> C Explain why titanium and aluminium are expensive.

Reduce, reuse, recycle

Many people recycle aluminium cans. Creating new cans from aluminium uses only about 10% of the energy needed to make cans from newly extracted bauxite. Less pollution is created, too.

Questions

1. Explain which properties of titanium make it suitable for artificial hips.
2. Explain why aluminium cannot be extracted from its ore by heating the ore with carbon.
3. List and explain two advantages of recycling aluminium.
4. Use ideas about particles and properties to explain why aluminium alloys are used in aeroplanes, not the pure metal.

E
↓
C
↓
A*

▲ A bauxite mine

▲ Titanium ore

Did you know...?

Titanium artificial joints are so resistant to corrosion that they can stay in place for 20 years.

Exam tip   AQA

✓ Aluminium and titanium are useful because they have low densities and they do not corrode. They are expensive because they are extracted from their ores in multi-step processes which need lots of energy.

Learning objectives

After studying this topic, you should be able to:

✔ describe what crude oil is

✔ explain how oil companies get useful fractions from crude oil

A Name three fossil fuels.

B Explain why oil is a finite resource.

Limited supplies

Have you washed, got dressed, or travelled today? If so, you have probably used products made from **crude oil**, like those pictured here. Crude oil is a vital fuel and raw material. Oil companies extract millions of tonnes of it every day.

Crude oil is a **fossil fuel**. It was formed from the decay of buried dead sea creatures. The process took millions of years. We use up oil faster than new oil forms. So oil is **non-renewable**.

Oil wasn't made just anywhere. The conditions had to be exactly right. So today's reserves are limited, and will run out one day. Oil is a **finite** resource.

Coal and natural gas are fossil fuels too – it took millions of years to make them. Coal was formed from trees that were buried under swamps. Natural gas, like oil, was formed from dead sea creatures.

What's in crude oil?

Crude oil contains many different **hydrocarbons**. Hydrocarbons are compounds made up of hydrogen and carbon only.

Crude oil is a **mixture**. This means its different hydrocarbon compounds are not chemically joined together. Each of the hydrocarbons has its own properties. Being part of a mixture does not affect these properties.

Did you know...?

One of the biggest oil refineries in the world is at Jamnagar in India. It processes more than $100\,000\ \text{m}^3$ of crude oil a day – enough to fill 40 Olympic size swimming pools.

You can separate mixtures by physical means:

- Filtration separates a solid from a liquid.
- Distillation separates liquids with different boiling points.

C Explain the meaning of the word hydrocarbon.

D Give two characteristics of a mixture.

Fractions

Crude oil is not much use as it is. But separate it into **fractions** and you get valuable fuels and raw materials. A fraction is a mixture of hydrocarbons with similar numbers of carbon atoms and similar boiling points.

Separating oil fractions

Oil companies use the property of boiling point to separate crude oil into useful fractions by **fractional distillation**. The process is continuous – it carries on all the time.

Fractional distillation involves heating crude oil to about 450 °C. Its compounds **evaporate** to become gases. The gases enter the bottom of a **fractionating column**. The column has a temperature gradient. It is hot at the bottom and cooler at the top.

The gases move up the column. As they move up, they cool down. Different fractions **condense** to form liquids again at different levels.

- Compounds with the highest boiling points condense at the bottom of the column, and leave as liquids.
- Lower boiling point compounds condense higher up, where it is cooler, and leave as liquids.
- The lowest boiling point compounds leave from the top, as gases.

Questions

1 Give two characteristics of fossil fuels.

2 What is a crude oil fraction? Name three fractions, and give their uses.

3 Describe how crude oil is separated into fractions by fractional distillation. Use the words boil, evaporate, and condense in your answer.

Key words

crude oil, fossil fuel, non-renewable, finite, hydrocarbon, mixture, fraction, fractional distillation, evaporate, fractionating column, condense

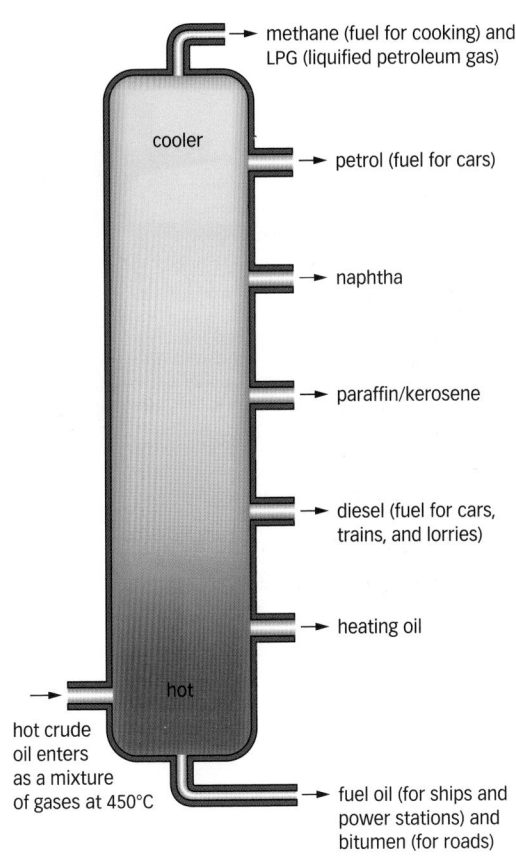

▲ Fractionating column

Exam tip AQA

✓ Remember – high boiling temperature fractions are removed at the bottom of a fractionating column; lower boiling temperature fractions come out higher up. You do not need to remember the names of specific oil fractions.

Learning objectives

After studying this topic, you should be able to:

✔ use molecular and displayed formulae to represent alkanes

✔ know the formulae of the first three alkanes

 Autogas is a form of LPG

Exam tip AQA

✔ Remember the patterns in properties of alkanes to help you answer questions about their suitability for use as fuels.

✔ Learn the formulae of methane, ethane, propane, and butane. You do not need to learn the formulae or names of other alkanes.

Inside oil fractions

Some cars are fuelled by LPG, liquefied petroleum gas. Their exhaust emissions are cleaner than those from diesel cars.

Liquefied petroleum gas is obtained from crude oil. It is a mixture of two hydrocarbons with similar boiling points, propane and butane.

A propane molecule consists of three carbon atoms joined to eight hydrogen atoms. Its **molecular formula** is C_3H_8. You can also represent propane by its **displayed formula**. This shows how the atoms are arranged in its molecules. Each letter represents an atom. Each line represents a covalent bond between two atoms.

Butane molecules are made up of four carbon atoms joined to ten hydrogen atoms. The molecular formula of butane is C_4H_{10}. Its displayed formula is

Propane and butane are **alkanes**. Alkanes are **saturated** hydrocarbons. This means their carbon atoms are joined together by single covalent bonds. Most other hydrocarbons in crude oil are alkanes, too.

- Methane is the main compound in natural gas. Its formula is CH_4.
- Ethane is also a gas. Its formula is C_2H_6.
- Hexane is part of the petrol fraction. Its formula is C_6H_{14}.

> A Draw displayed formulae for methane and ethane.
>
> B Give the numbers of carbon and hydrogen atoms in a molecule of hexane.

All alkanes have the same **general formula**, C_nH_{2n+2}. This means that the number of hydrogen atoms in an alkane is twice the number of carbon atoms plus two.

Size matters

Th combustion of hydrocarbon fuels releases energy. The uses of alkanes as fuels depend on their properties. And their properties depend on the sizes of their molecules.

Viscosity

Hexadecane molecules are long. They get tangled up. So hexadecane is difficult to pour and does not flow easily. It is a **viscous** liquid at room temperature. Pentane is also liquid at room temperature. Its smaller molecules make it runnier – or less viscous – than hexadecane. There is a pattern in the viscosity of liquid alkanes – the longer the hydrocarbon chain, the more viscous the liquid.

Melting and boiling points

Molecule size also influences boiling points. Again, there is a pattern – the smaller the molecule, the lower the boiling point.

Number of carbon atoms in hydrocarbon chain	State at room temperature
1 to 4 (smaller molecule)	gas
5 to 16	liquid
17 or more (larger molecule)	solid

Igniting alkanes

Alkanes with small molecules catch fire more easily than those with bigger molecules.

▲ Methane is a good fuel for cooking because it ignites easily

Key words

molecular formula, displayed formula, alkane, saturated, general formula, viscous

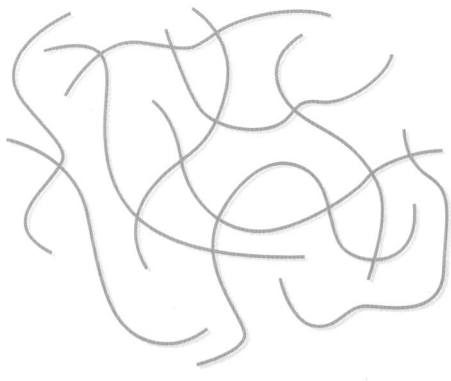
▲ Longer chain alkanes are more viscous because their molecules get more tangled

C Describe the properties of methane that make it a good fuel for cooking.

D Explain why the properties of hexadecane make it a suitable vehicle fuel when mixed with other hydrocarbons.

Questions

1. Give the numbers of carbon and hydrogen atoms in a molecule of pentane, C_5H_{12}. ↓ E

2. Draw a displayed formula for pentane.

3. Use the general formula for alkanes to work out the formula of an alkane with 15 carbon atoms. ↓ C

4. Explain why an alkane with molecule formula $C_{18}H_{38}$ is not an ideal cooking fuel. ↓ A*

Learning objectives

After studying this topic, you should be able to:

- ✔ identify hydrocarbon combustion products
- ✔ explain their impacts on the environment

▲ Fire dancers burn kerosene on their firesticks

▲ Didcot power station burns coal and gas

▲ These trees have been damaged by acid rain

Dancing with fire

Aaden is a fire dancer. Backstage, he dips the wick of his firesticks into kerosene and sets them alight. Then he leaps onto the stage and thrills his audience with spectacular displays of twirling fire and hot dance moves.

Kerosene is a mixture of hydrocarbons. When it burns outside, there is plenty of oxygen around. Its combustion makes mainly carbon dioxide gas and water vapour. For example:

undecane + oxygen → carbon dioxide + water

$$C_{11}H_{24} + 17O_2 \rightarrow 11CO_2 + 12H_2O$$

This is an example of **complete combustion**.

Greenhouse gas

Carbon dioxide is a greenhouse gas. Find out more about its impacts on the environment on spread C1.16.

Killer rain

Didcot power station burns coal and gas to generate electricity. Coal contains sulfur impurities. So when coal burns it produces **sulfur dioxide** gas as well as carbon dioxide and water. If sulfur dioxide goes into the air, it dissolves in water in clouds. This makes **acid rain** – rain which is more acidic than normal.

Oxides of nitrogen such as nitrogen dioxide, NO_2, are formed when hydrocarbon fuels burn at high temperatures in car engines. These also dissolve in water to make acid rain.

Acid rain has many environmental impacts:

- It makes lakes more acidic. Some species of water animals and plants cannot survive if the water is too acidic.
- Acid rain damages trees. It dissolves soil nutrients and washes them away before tree roots can take them in. It also damages the protective waxy coating of leaves. The leaves can no longer produce enough food for the tree.
- Acid rain damages limestone buildings by reacting with the calcium carbonate of the limestone (see spread C1.6).

There are signs that the UK is beating acid rain. Scientists have discovered how to remove sulfur impurities from fuels such as diesel.

It is less easy to remove sulfur impurities from coal. So some power stations remove sulfur dioxide from their waste gases by adding limestone powder to the gases. There is a chemical reaction. Calcium carbonate in limestone reacts with sulfur dioxide to make calcium sulfate. Calcium sulfate is a useful product. It is used to make plasterboard for houses.

> A Name the gases that cause acid rain.
>
> B Describe and explain three impacts of acid rain.
>
> C Describe two ways of reducing acid rain.

Particulate problem

This car runs on diesel fuel. Burning diesel produces tiny pieces of solid, or **particulates**, as well as carbon dioxide and water. Particulates may be made up of soot, a form of carbon, and unburned fuels.

Particulates may cause **global dimming**. In the atmosphere, particulates reflect sunlight back into space. As more particulates enter the atmosphere, less sunlight reaches the Earth's surface.

Silent killer

Edith heats her house with a gas boiler. She hasn't had it checked recently. One day, her neighbour finds her confused, drowsy, and barely able to breathe. Edith has **carbon monoxide** poisoning.

The killer gas was produced because too little air was reaching the flame of the boiler. **Incomplete combustion**, or **partial combustion**, happened, so carbon monoxide, CO, was formed as well as carbon dioxide.

Key words

complete combustion, sulfur dioxide, acid rain, particulates, global dimming, carbon monoxide, incomplete combustion, partial combustion

Exam tip AQA

✔ Remember – sulfur dioxide and oxides of nitrogen cause acid rain, particulates cause global dimming, and carbon monoxide is poisonous.

> D Describe an environmental impact of particulates.

Questions

1 Give two symptoms of carbon monoxide poisoning. ↓ E

2 Explain the difference between complete and incomplete combustion. ↓ C

3 Make a table to summarise the environmental impacts of three products of combustion of hydrocarbons. ↓ A*

Learning objectives

After studying this topic, you should be able to:

✔ describe and explain the impacts of the greenhouse gas carbon dioxide

Key words

greenhouse gas, global warming, climate change, alternative fuel, carbon neutral

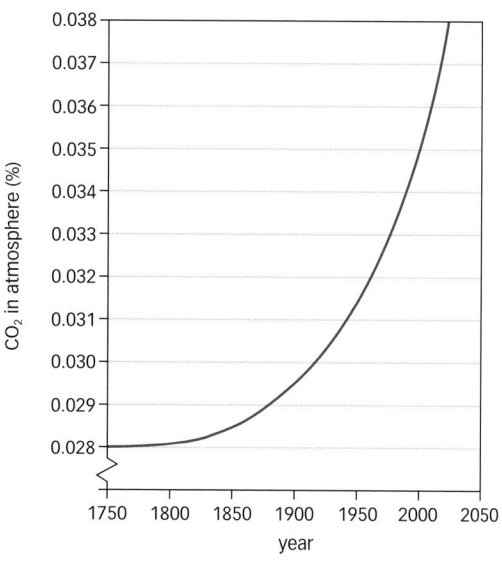

▲ Graph of carbon dioxide concentration over time

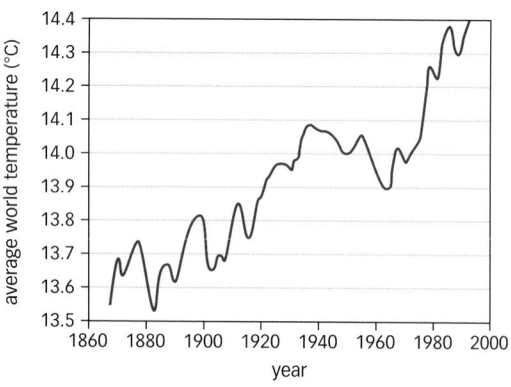

▲ Graph of mean average world temperature over time

Greenhouse gas

Burning fossil fuels produce carbon dioxide. Carbon dioxide is a **greenhouse gas**. Its presence in the atmosphere helps keep Earth warm enough for life.

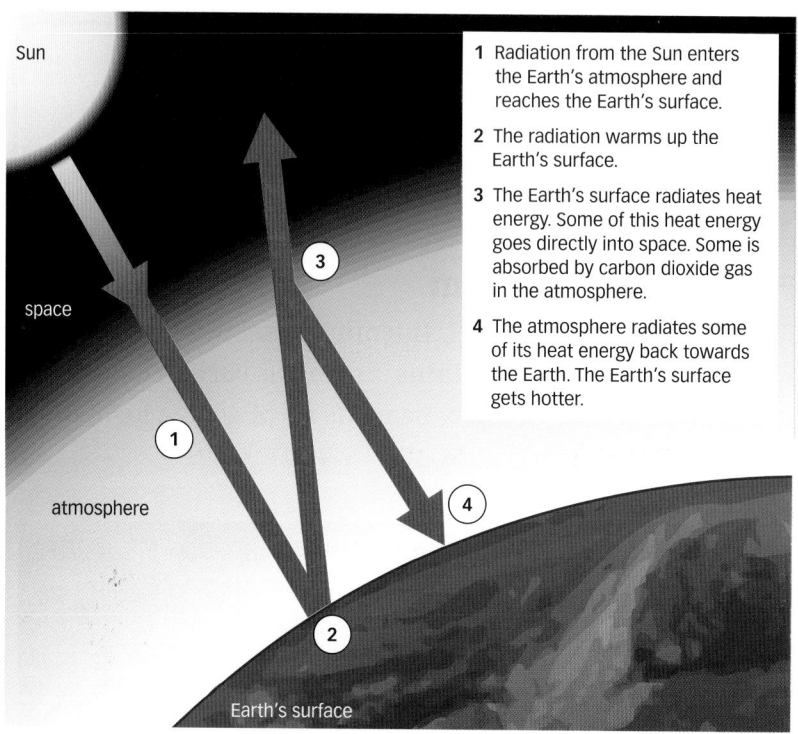

1 Radiation from the Sun enters the Earth's atmosphere and reaches the Earth's surface.

2 The radiation warms up the Earth's surface.

3 The Earth's surface radiates heat energy. Some of this heat energy goes directly into space. Some is absorbed by carbon dioxide gas in the atmosphere.

4 The atmosphere radiates some of its heat energy back towards the Earth. The Earth's surface gets hotter.

▲ Global warming

Global warming – the cause

Over time, the average air temperature of the Earth has increased. This is **global warming**. At the same time, the percentage of carbon dioxide in the atmosphere has also increased. The graphs show these changes.

In the 1950s, scientists wondered if extra carbon dioxide might be causing the temperature increase. They began collecting evidence. Since the 1980s, hundreds of scientists have researched global warming. There is now a huge body of evidence supporting the theory that extra carbon dioxide from human activities causes global warming.

A Describe how greenhouse gases keep the Earth warm.

B Describe and explain what the graphs on the left show.

Global warming – the impacts

The impacts of global warming include:

- **Climate change** – the Earth's weather patterns are changing. Some people are already dealing with more extreme weather events. Some areas will have more droughts. Others will have more flooding. Climate change puts some plant and animal species at risk of extinction. Other species – such as mosquitoes that carry the deadly disease of malaria – may spread over wider areas.
- Melting ice caps – higher average temperatures are making polar ice caps melt. Sea levels are rising and coastal areas – including big cities – are at risk of flooding.

> C Describe and explain two impacts of global warming.

Global warming – some solutions

Using less

Governments, councils, and environment groups are urging people to consume less and use less fuel.

Alternative fuels

Scientists are playing their part in trying to prevent climate catastrophe. They have developed cars fuelled by **alternative fuels**, such as hydrogen and ethanol.

Hydrogen cars produce one exhaust product – water vapour. But there are problems. Hydrogen fuel must be manufactured – either from methane or by using electricity to break up water molecules. Both these processes produce carbon dioxide gas. Storing and transporting hydrogen fuel – which is an explosive gas – are difficult and expensive.

In Brazil, many cars are fuelled by ethanol. The ethanol is made from crops such as sugar cane. Some people say that fuel produced from renewable crops is **carbon neutral**. Growing sugar cane takes carbon dioxide from the atmosphere. Burning ethanol returns a similar amount of carbon dioxide to the atmosphere. But the balance is not perfect. Energy is needed to make fertilisers and to manufacture ethanol. Both processes produce carbon dioxide gas. Some people think that it is unethical to use land to grow fuel instead of food.

▲ Climate change causes flooding

Exam tip   **AQA**

✓ Make sure you know the difference between the terms 'global warming', 'greenhouse gas', and 'climate change'.

▲ Ethanol fuel can be produced from sugar cane

Questions

1 Name a fuel produced from sugar cane.

2 Some people say that ethanol produced from sugar cane is carbon neutral. Explain why.

3 Draw and annotate a big diagram to explain the causes and effects of global warming.

4 Describe and explain the problems associated with producing and using hydrogen and ethanol fuels.

Learning objectives

After studying this topic, you should be able to:

✔ evaluate the benefits, drawbacks, and risks of using plant oils to produce fuels

A Explain why plant oils make good fuels.

B Use the data to identify which of the three plant oils transfers the most energy on burning.

C Suggest why plant oils are used for fuel even though they transfer less energy than diesel.

▲ Oil palms produce a useful fuel

Plant oil fuels

Brendan has a diesel car. But he no longer buys expensive diesel fuel. Instead, he produces cheap fuel in his garage – from used cooking oil from the chip shop down the road. Brendan had to convert his car engine to prevent damage from the cooking oil fuel.

◀ Biofuel for cars

Plant oils make good fuels for cars, buses, and trains because they transfer large amounts of energy when they burn. The table compares plant oils to diesel.

Fuel	Energy transferred on burning 1 kg of the fuel (kJ, approximate values)
diesel	45 000
sunflower oil	38 000
peanut oil	40 000
rapeseed oil	37 000

Plant oils are not the only type of **biofuels** suitable for use in vehicles. Two other types are:

- ethanol produced from crops such as sugar cane (see spread C1.16)
- biodiesel produced by chemically reacting plant oils or animal fats with an alcohol.

Palm oil palaver

In 2009, a company asked for permission to build a biofuel power station in Bristol. The power station would burn oil from oil palm trees as well as other plant oils. The burning reaction would heat water to make steam. The steam would be used to turn turbines. The turbines would generate enough electricity for 25 000 homes.

Many people expressed their opinions about the power station.

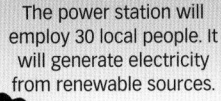

The power station will employ 30 local people. It will generate electricity from renewable sources.

It is nonsense to say that palm oil fuel is a leading cause for rainforest destruction.

In countries like Indonesia, tropical rainforests are destroyed to make space to grow oil palms. This damages biodiversity and has decimated orang-utan and elephant populations.

Burning plant oils results in emissions of oxides of nitrogen gases. These worsen asthma and other lung diseases.

One in six people in the world do not have enough to eat. It is immoral to use land to grow plants for fuel and not for food.

Oil palms remove carbon dioxide from the atmosphere as they grow. Similar amounts of carbon dioxide are returned to the atmosphere by burning palm oil.

A palm oil company evicted us from our land. We have nowhere to go.

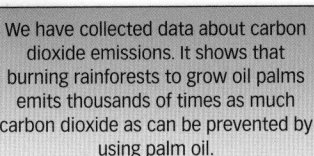

We have collected data about carbon dioxide emissions. It shows that burning rainforests to grow oil palms emits thousands of times as much carbon dioxide as can be prevented by using palm oil.

In 2010, Bristol City Council decided not to give the go-ahead for the biofuel power station.

D Identify two benefits of using plant oils to generate electricity.

Key words

plant oils, biofuel

Questions

1 Identify two ethical and two environmental arguments against using plant oils to generate electricity.

2 Explain two drawbacks of using palm oil as a fuel.

3 Suggest two reasons for Bristol City Council not allowing the power station to be built.

4 Write a paragraph to describe and explain the benefits, drawbacks, and risks of using plant oils as fuels.

Exam tip AQA

✓ Plant oil fuels are renewable. However, they produce carbon dioxide on burning. Growing plants for oil may lead to environmental destruction.

Course catch-up

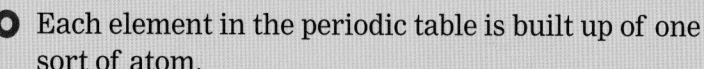

Revision checklist

- Each element in the periodic table is built up of one sort of atom.
- Atoms have a nucleus containing protons and neutrons surrounded by electrons within energy levels (shells).
- Elements in the same group of the periodic table have the same number of electrons in the highest energy level (outer shell). They react in similar ways.
- Balanced equations describe chemical reactions.
- Limestone (calcium carbonate) is a sedimentary rock and is quarried from the ground for use as a building material.
- Limestone is heated with clay to form cement. Cement, water, and sand form concrete.
- Metal carbonates are decomposed by heat to form metal oxides + carbon dioxide.
- Calcium oxide reacts with water to form calcium hydroxide. This reacts with carbon dioxide to reform calcium carbonate.
- The elements in the middle block of the periodic table are transition metals.
- Pure iron is converted into an alloy (steel) by mixing with other elements. This alters the arrangement of atoms.
- Metals are extracted from metal ores by electrolysis, heating, or reaction with carbon (reduction).
- Aluminium and titanium have low density and do not corrode. They are extracted using expensive electrolysis.
- Recycling metals saves resources, energy, and waste.
- Crude oil is a non-renewable resource. It is a mixture of hydrocarbons (separated by fractional distillation).
- Alkanes are saturated hydrocarbons which contain only single bonds and have general formula C_nH_{2n+2}.
- Burning hydrocarbon fuels releases carbon dioxide (greenhouse effect), carbon monoxide (toxic), sulfur oxides and nitrogen oxides (acid rain), particulates (global dimming).
- Biofuels (biodiesel, bioethanol) are produced from plant material. Producing biofuels releases less CO_2 overall and uses renewable resources. It may reduce the amount of food crops the world can grow.

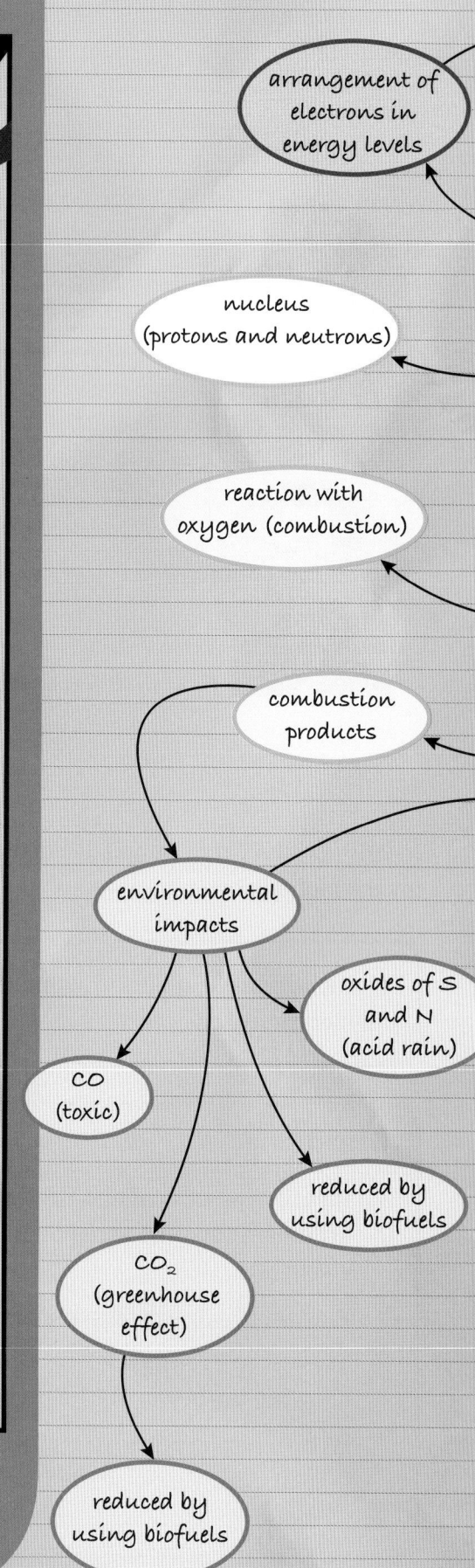

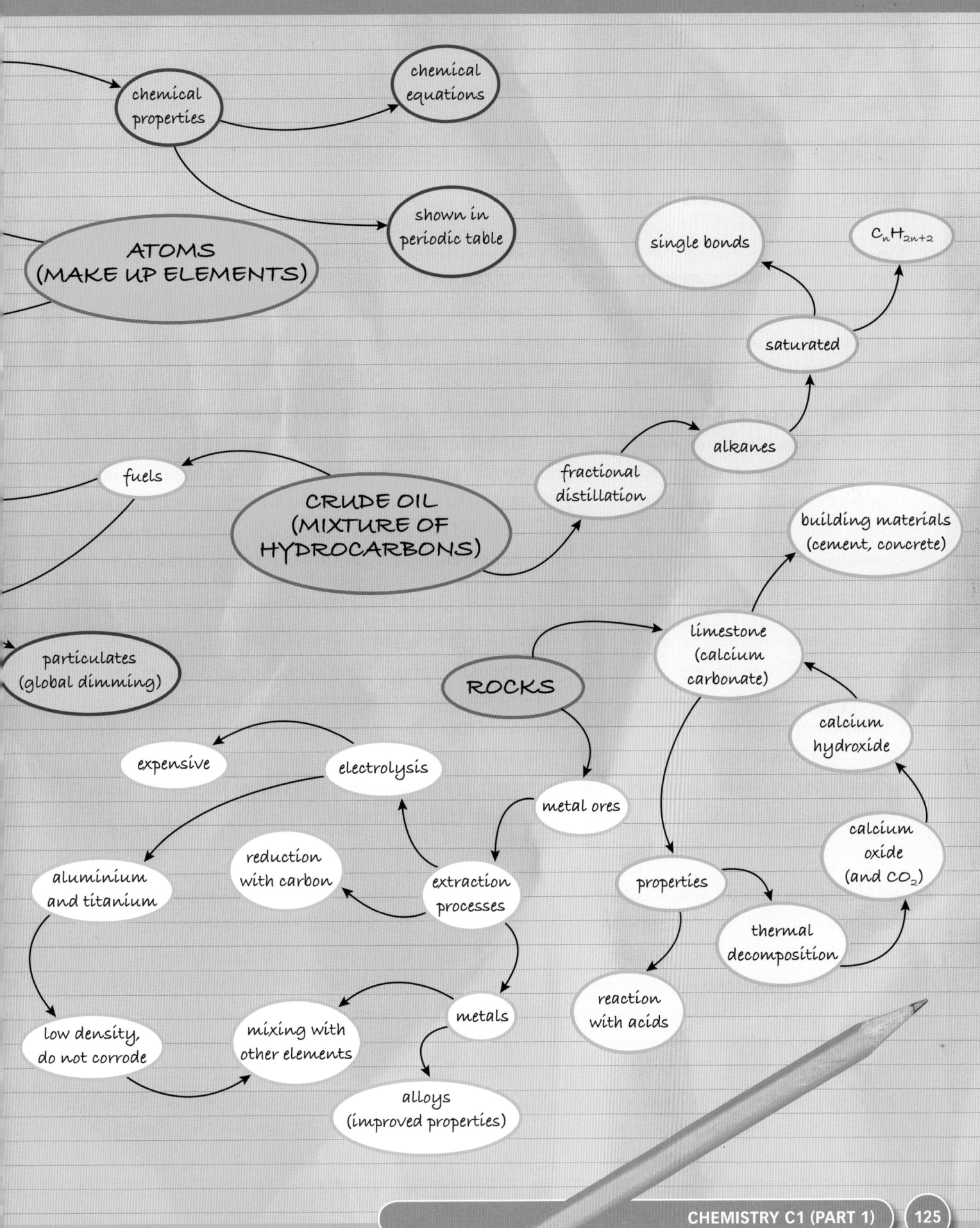

chemical properties

chemical equations

shown in periodic table

ATOMS (MAKE UP ELEMENTS)

single bonds

C_nH_{2n+2}

saturated

alkanes

fuels

CRUDE OIL (MIXTURE OF HYDROCARBONS)

fractional distillation

building materials (cement, concrete)

limestone (calcium carbonate)

particulates (global dimming)

ROCKS

calcium hydroxide

expensive

electrolysis

metal ores

calcium oxide (and CO_2)

aluminium and titanium

reduction with carbon

extraction processes

properties

thermal decomposition

low density, do not corrode

mixing with other elements

metals

reaction with acids

alloys (improved properties)

AQA Upgrade

Answering Extended Writing questions

Many drinks cans are made of aluminium. In 2008, people in Germany recycled 96% of their aluminium cans, while just 51% of British aluminium cans were recycled. The rest were buried in landfill sites.

Give reasons why Britain should increase the percentage of aluminium cans it recycles.

The quality of written communication will be assessed in your answer to this question.

If you resycle alyouminium cans there will be less climate warming. If you throw away cans on the street, they mite hurt squirrels and birds. Britain should resycle as much as Germany.

G–E

Examiner: The candidate knows that recycling aluminium results in less global warming than extracting the metal from its ore, but has not explained why. There is no scientific detail, and the candidate has muddled two scientific terms (global warming and climate change). There are several spelling errors.

Recycling aluminium is better than getting it from its rock because recycling makes less polution! Recycling also needs less energy, and makes less carbon diokside! Getting aluminium from its rock causes climate change!

D–C

Examiner: The answer includes four points explaining the advantages of recycling aluminium. The candidate uses several scientific words correctly, but uses the word 'rock' instead of 'ore'. The answer is well structured, but includes spelling and punctuation errors. The answer would be improved by explaining the links between energy, carbon dioxide, and climate change.

Recycling more aluminium means that less aluminium needs to be produced from its ore, which is good since the ore is a finite resource.
Extracting aluminium from its ore makes dangerous pollution. Recycling aluminium does not.
Electricity is needed to get aluminium from its ore. Generating electricity makes carbon dioxide, which causes global warming. Recycling aluminium needs much less energy.
Aluminium cans that are not recycled go to ugly landfill sites.

B–A*

Examiner: The answer clearly explains the scientific reasons for recycling aluminium. The points are well explained and in a sensible order. The answer includes scientific words, used correctly. The spelling, punctuation, and grammar are accurate. The answer would be improved by mentioning that it is electricity generated from fossil fuels that causes carbon dioxide.

Exam-style questions

1 Choose metals from this list to answer the questions about the ways in which metals are extracted from their ores:

gold   iron   aluminium

A01 **a** Which metal is extracted from its ore in the blast furnace?

A01 **b** Which metal is extracted by electrolysis?

A01 **c** Which metal is so unreactive that it is found naturally in the Earth's crust?

2 Steel is an alloy made up of iron with added carbon. Adding carbon affects the properties of steel. This graph shows the results of an experiment done to find out how the strength of steel is affected by the amount of carbon added.

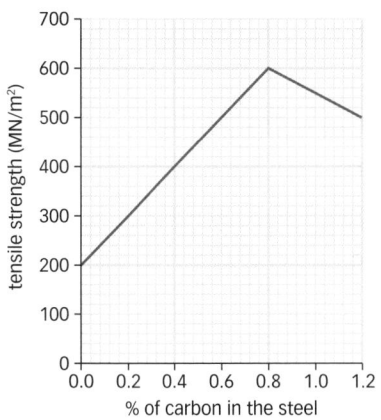

A03 **a** How does increasing carbon percentage in the alloy affect its strength?

A03 **b** Imagine that steel with a tensile strength of 500 MN/m² needs to be produced. Give two possible percentages of carbon which will produce a suitable steel.

A01 **c** Give one other property of metals, apart from strength, which can be improved by using an alloy.

3 A sodium atom has 11 electrons. It also contains protons and neutrons.

A02 **a** Draw a labelled diagram to show how these particles are arranged in an atom of sodium.

A01 **b** Give an example of a chemical property which is similar for all Group 1 elements.

Extended Writing

4 **A01** Limestone is a rock found in some parts of the UK. Large quarries are dug to get the limestone out of the ground. Why is limestone an important resource, and why do some people object to limestone quarries?

5 **A02** Metal carbonates, such as limestone, decompose into carbon dioxide and metal oxides when heated. Some carbonates decompose at a higher temperature. Describe an experiment to find out which of two carbonates decomposes at the highest temperature.

6 **A02** **A03** Biodiesel is an example of a biofuel. It can be produced from plant oils such as palms or rapeseed. Scientists debate whether the use of biofuels is a benefit to the environment. Discuss the arguments for and against the use of biofuels.

G–E

D–C

B–A*

A01   Recall the science

A02   Apply your knowledge

A03   Evaluate and analyse the evidence

C1 Part 2

Polymers, plant oils, the Earth, and its atmosphere

Why study this unit?

Every year, volcanoes, and earthquakes cause death and destruction. But what causes these disasters? Can scientists predict them and so save lives?

In this unit you will discover how gigantic pieces of the Earth's crust grind against each other to cause earthquakes, and find out how scientists monitor volcanoes to predict eruptions.

As well as providing fuel, crude oil provides most of the huge variety of plastics we use every day. How are these plastics made? How can we minimise their impact on the environment?

You will learn how chemicals from crude oil make plastics, and about their renewable replacements. You will also consider how burning fossil fuels changes our atmosphere, leading to acid rain and global warming.

You should remember

1   In chemical reactions, atoms are rearranged to make new substances.

2   There are patterns in the chemical reactions of substances.

3   The way we use materials depends on their properties.

4   There are patterns in the properties of materials.

5   Humans extract many useful materials from plants.

6   The Earth consists of three layers – the crust, mantle, and core.

7   Earthquakes and volcanoes occur mainly in certain regions of the world.

8   The Earth's atmosphere is a mixture of gases.

The 1816 volcanic eruption of Mount Tambora in Indonesia killed up to 92 000 people, including 10 000 from the explosion and ash fall, and 82 000 from other causes. The year 1816 became known as 'the year without a summer', because volcanic ash in the atmosphere lowered global temperatures by an average of between 0.4 °C and 0.7 °C. Lower temperatures led to failed harvests and severe food shortages in many parts of the world.

This photo shows the July 1980 eruption of Mount St Helens, Washington

Learning objectives

After studying this topic, you should be able to:

- explain how cracking makes useful products
- identify the products of cracking reactions
- identify an economic benefit of cracking reactions

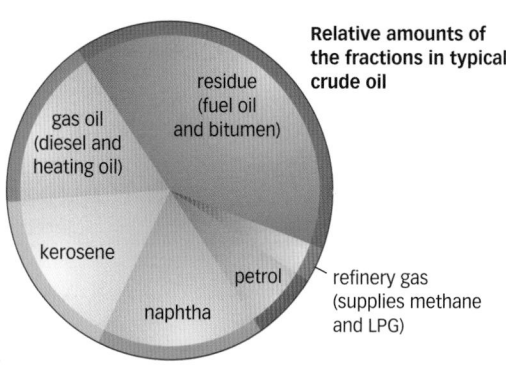

Relative amounts of the fractions in typical crude oil

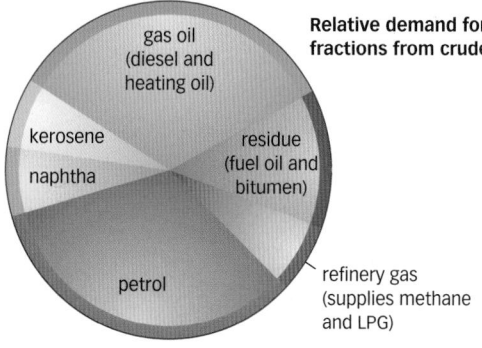

Relative demand for fractions from crude

▲ These charts compare the supply and demand for crude oil fractions

▲ A catalytic cracking tower at an oil refinery

Petrol, plastics, and propanone

What do the things in the pictures have in common? Petrol, polythene bottles, and propanone for nail varnish remover are all made from chemicals obtained from crude oil. Thousands of other important chemicals come from crude oil too.

Cracking

The fractional distillation of crude oil makes useful products, including petrol. But the supply of some fractions does not match demand (the amounts people need).

Companies use **cracking** to break down bigger hydrocarbon molecules to smaller, more useful molecules.

Cracking involves heating an oil fraction to a high temperature. The hydrocarbons of the fraction vaporise. Then the vapour passes over a hot **catalyst**. Alkane molecules in the vapour break down to form smaller molecules in thermal decomposition reactions. A catalyst is a substance that speeds up the reaction. It is not used up in the reaction itself. Hydrocarbons can also be cracked by mixing their vapours with steam and heating to a very high temperature.

Octane has the formula C_8H_{18}. It is an alkane in the naphtha fraction. In cracking, each octane molecule breaks down to make two smaller molecules, such as hexane (C_6H_{14}) and ethene (C_2H_4). The equation for this cracking reaction is

$$C_8H_{18} \rightarrow C_6H_{14} + C_2H_4$$

Hexane is an alkane. It is a useful fuel. Hexane is added to the petrol fraction, so the company now has more petrol to sell.

Inside ethene

Ethene is a hydrocarbon. It is made up of atoms of carbon and hydrogen only.

You can represent ethene by its molecular formula, C_2H_4. Its displayed formula shows how its atoms are joined together.

$$\underset{H}{\overset{H}{\diagdown}}C=C\underset{H}{\overset{H}{\diagup}}$$

The double line between the two carbon atoms represents a double covalent bond. Double bonds are stronger than single bonds. Because ethene has a double bond it is an **unsaturated hydrocarbon**. The double bond makes ethene more reactive than ethane.

You can detect compounds with double bonds by testing with a solution of bromine, or **bromine water**. Orange bromine water becomes colourless when it reacts with ethene and other unsaturated compounds.

bromine water — | — bromine water after ethene has been bubbled through it

The alkene family

Ethene belongs to a family of hydrocarbons called the **alkenes**. All alkenes are unsaturated. Another alkene is propene, C_3H_6

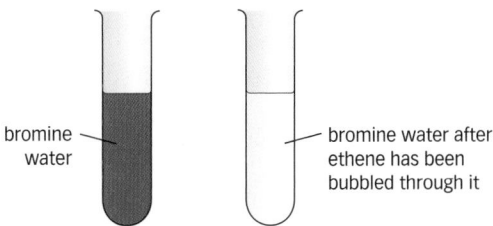

Alkenes have the general formula C_nH_{2n}. This shows that the number of hydrogen atoms in an alkene molecule is double the number of carbon atoms.

C Which family of hydrocarbons is propene from?

D A butene molecule has four carbon atoms. Use the general formula of alkenes to predict the number of hydrogen atoms in the molecule.

Key words

cracking, catalyst, unsaturated hydrocarbon, bromine water, alkene

A Identify the conditions needed for cracking reactions.

B Give an economic benefit to oil companies of cracking reactions.

Exam tip

✔ Remember – cracking breaks down big molecules to produce smaller molecules of alkanes and alkenes.

Questions

1 Name two possible products produced by cracking octane. ↓ E

2 What is a catalyst?

3 Cracking produces hydrocarbons of two families. Name these families and give their general formulae. ↓ C

4 A cracking reaction produces pentane (C_5H_{12}) and ethene (C_2H_4). Predict the formula of the starting material. ↓ A*

5 Suggest an environmental disadvantage of making products from chemicals obtained from crude oil.

Learning objectives

After studying this topic, you should be able to:

✔ describe how polymers are made from monomers

✔ evaluate the social impacts of using polymers

▲ Many items we use every day are made of plastics

A List six items that contain plastics.

B Define the words polymer and monomer.

Did you know...?

The polymers poly(ethene), PVC, and Teflon were all discovered by accident.

Plastics everywhere

Look around you. How many things are made of plastics? Imagine life without plastics. Your great-great-grandparents probably did live without most of them – plastics only started to be widely used in the 1930s.

Most of the things we call plastics are made from polymers. But what's in a polymer?

Polymers are materials that have very big molecules. They are made by joining together thousands of small molecules, called **monomers**.

Inside polythene

Polythene is an important polymer. Its properties make it useful. It is strong, flexible, and durable – perfect for bags and bottles.

Polythene molecules consist of thousands of atoms of carbon and hydrogen. The atoms are joined together in long chains.

The structure of polythene explains its properties:

- It is strong because the atoms in a molecule are joined together tightly, so it is difficult to break up a molecule.
- It is flexible because its molecules can slide over each other.

Making polythene

Polythene molecules are made by joining together thousands of ethene molecules. Ethene molecules can join together because they have a double bond. The diagram shows how the molecules join together. Only a small part of the polythene molecule is shown.

part of a poly(ethene) molecule

▲ Scientists call polythene **poly(ethene)**. The name shows that it is made from many – or poly – ethene molecules.

This is an example of a **polymerisation reaction**. You can use beads, paper clips, or molecular model kits to model polymerisation reactions.

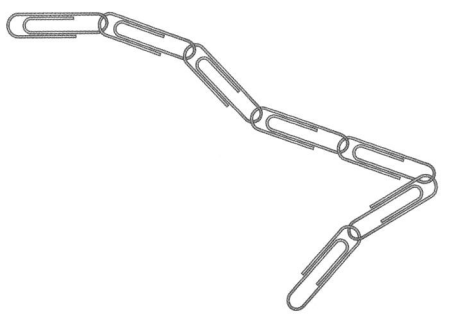

◄ Each paper clip represents a monomer molecule. The chain represents the polymer.

More polymers

Poly(ethene) is not the only polymer. There are thousands of others. One of these is **poly(propene)**.

Poly(propene) is strong and rigid. It is not damaged by high temperatures. You can bend poly(propene) lots of times without it breaking. These properties mean that poly(propene) is useful for making many things, including:

- ropes
- underground water pipes
- dishwasher-safe food containers
- hinges for flip-top bottles.

The monomer used to make poly(propene) is propene. The formula of propene is C_3H_6. Its atoms are joined together like this:

Thousands of propene molecules join together in long chains to make poly(propene). You can represent the reaction like this:

part of a poly(propene) molecule

Key words

polymer, monomer, poly(ethene), polymerisation reaction, poly(propene)

C Match each of the uses of poly(propene) listed on this page to one or more properties that make it suitable for this purpose.

D Name the monomer used to make poly(propene).

Exam tip **AQA**

✔ Practise writing equations to show how a polymer such as poly(propene) is made from its monomer.

Questions

1 Name the monomer and polymer in the polymerisation reaction of ethene. ↓ E

2 Explain why alkene molecules can join together.

3 Write an equation to show how poly(propene) is formed from its monomer. ↓ C

4 Describe three benefits to people of the invention of polymers. These are social benefits.

5 Describe how you could use people to model a polymerisation reaction. ↓ A*

Learning objectives

After studying this topic, you should be able to:

✔ explain how the properties of polymers depend on how they are made

✔ explain how the properties of polymers determine their uses

▲ HDPE makes rigid garden furniture

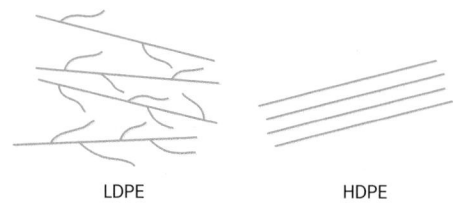

LDPE   HDPE

▲ LDPE is less dense than HDPE because HDPE has fewer branches on its chains.

> **A** Identify two properties that make HDPE more suitable than LDPE for making outdoor furniture.
>
> **B** Explain why HDPE has a higher density than LDPE.

Fit for purpose

Poly(ethene) and poly(propene) are just two of the many hundreds of synthetic polymers that chemists have created. Some polymers were discovered accidentally. Others were developed after hours of painstaking work in laboratories.

Each polymer has its own unique properties. The properties depend on

• what the polymer was made from
• the conditions under which it was made.

Different polymers have different uses. Their uses depend on their properties.

Two types of poly(ethene)

You've probably used poly(ethene) bags. But did you know that you can get poly(ethene) garden furniture, too?

There are two types of poly(ethene):
• low density poly(ethene), **LDPE**
• high density poly(ethene), **HDPE**.

Each type of poly(ethene) has its own properties.

	LDPE	HDPE
Density (g/cm³)	0.92	0.95
Maximum temperature at which the polymer can be used (°C)	85	120
Strength (megapascal, MPa)	12	31
Relative flexibility	flexible	stiff

LDPE and HDPE have different properties because their molecules have different structures.
• LDPE polymer molecules have side branches. The branches prevent the polymer molecules from lining up in a regular pattern, so the density is lower.
• HDPE polymer molecules have few side branches. Its molecules line up in a pattern, so the density is higher. The molecules are held together more strongly, so HDPE is stronger and has a higher melting point than LDPE.

Dental polymers

Do you have fillings in your teeth? For years, dentists have filled teeth with a mixture of mercury, silver, and other metals. But this material conducts heat well, making it uncomfortable to eat very hot or very cold food.

Now, your dentist may offer you a white **dental polymer** filling. He will put a paste into your tooth and shine ultraviolet light on it. The ultraviolet light starts a reaction in which polymer molecules in the paste join together to make a solid filling.

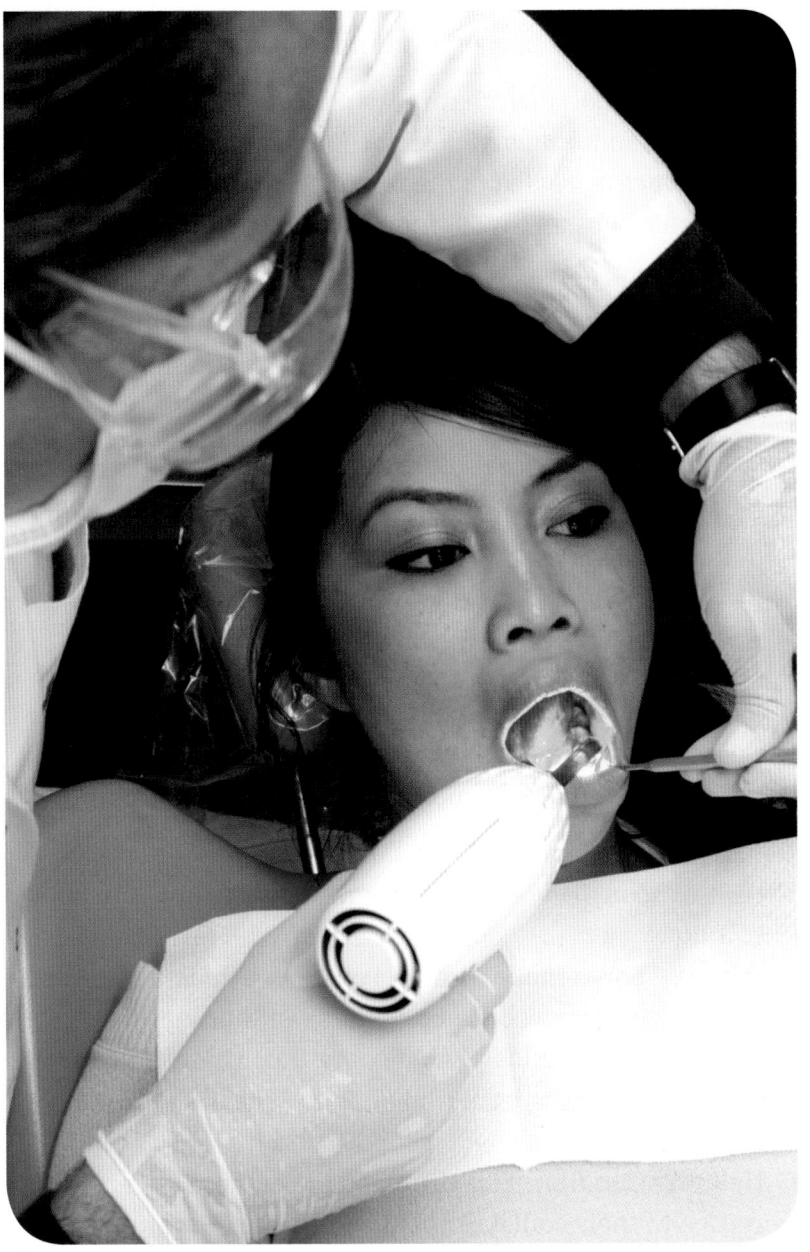

▲ Ultraviolet light starts a polymerisation reaction in the filling material

Key words

LDPE, HDPE, dental polymer

Exam tip

✓ Make sure you can explain how the properties of a polymer are linked to its uses. You do not need to remember details about specific polymers in your exams.

Questions

1 Explain why different polymers have different properties. E

2 List one use each for LDPE and HDPE. Explain how the properties of each of these polymers make them suitable for these uses. C

3 Suggest two advantages of filling your teeth with a white polymer and not a mixture of metals.

4 Use ideas about particles to explain why HDPE is stronger and has a higher melting point than LDPE. A*

Learning objectives

After studying this topic, you should be able to:

✔ explain how the properties of polymers determine their uses

▲ Disposable nappies contain hydrogels

▲ The monomer from which poly(sodium propenoate) is made.

B Explain why the properties of shrink wrap make it a suitable material for packaging items such as DVDs.

Key words

hydrogel, shape memory polymer, smart material, breathable material

Perfect polymers

Chemists continue to create new polymers with perfect properties for particular purposes. These include waterproof coatings for fabrics, wound dressings, and shape memory polymers.

Hydrogels

Disposable nappies absorb huge amounts of urine – and the baby doesn't even feel wet. How do they do this?

Disposable nappies contain **hydrogels**. Hydrogels are made from polymers such as poly(sodium propenoate).

Normally, the polymer chains are coiled up. But if you take away all the sodium particles, the chains uncoil. Water molecules are attracted to the uncoiled chains, and the hydrogel absorbs up to 500 times its own weight of water.

Hydrogels also make excellent wound dressings. The hydrogel protects the wound from infection and controls bleeding. Because hydrogels do not stick to the skin, you can remove them easily without damaging the skin.

A Explain why the properties of hydrogels make them suitable for use in disposable nappies.

Shape memory polymers

If you've bought a DVD recently, you might have noticed that it was shrink-wrapped. Shrink wrap is a **shape memory polymer**. Shape memory polymers are **smart materials** – they change in response to their environment.

Shrink wrap is made from polymers such as poly(ethene). Here's how:

- Heat the poly(ethene) until it becomes a thick liquid. The molecules are now coiled together randomly.
- Cool the liquid quickly. As it cools, quickly stretch out the polymer to make a thin film of solid. The molecules are now stretched out.
- Heat the thin film. The stretched molecules suddenly return to their coiled shape. The film shrinks and wraps tightly round the DVD case.

Waterproof clothing

Ben loves walking, but not in the rain. His waterproofs are made of nylon. Nylon stops water getting in. It is also tough and lightweight. But nylon stops water vapour getting out. So Ben's sweat condenses inside his waterproofs, making him feel damp and uncomfortable. Yuk!

Suzette's waterproofs are much more comfortable. They are made of a **breathable material**. The material contains three layers. The middle layer is made of a polymer called PTFE. This has many tiny holes. Each hole is too small for water droplets to pass through, but big enough for water vapour to pass through. So rain can't get in, but water vapour from sweat can get out.

▲ Breathable waterproof coats

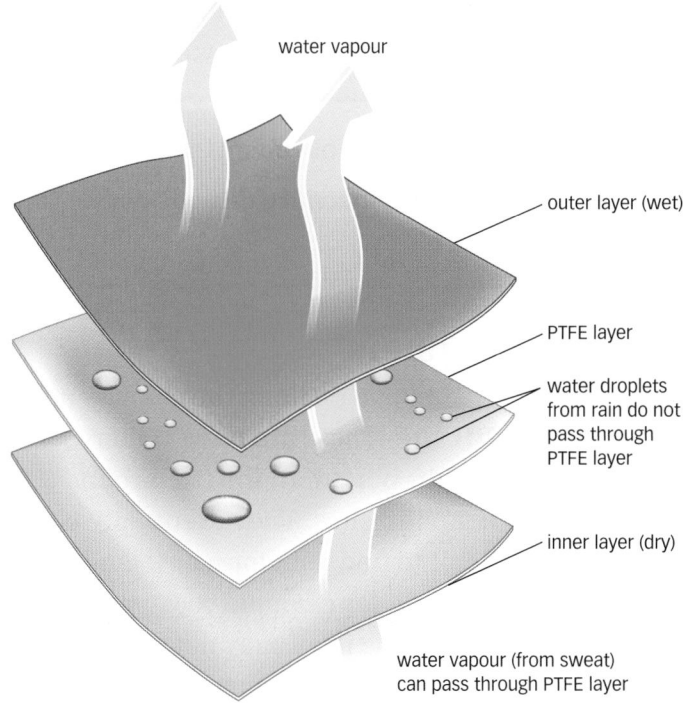

water vapour

outer layer (wet)

PTFE layer

water droplets from rain do not pass through PTFE layer

inner layer (dry)

water vapour (from sweat) can pass through PTFE layer

▲ Breathable fabrics are made up of materials with different properties

Did you know...?

The PTFE layer in breathable materials has around 1.4 billion holes per square centimetre.

C Explain why breathable materials stop water getting in but allow sweat to get out.

Questions

1 List two uses of hydrogels.

2 Give one use of shape memory polymers.

3 Make a table showing the uses of three types of polymer and explaining how their properties make them suitable for these purposes.

E

C

Exam tip   AQA

✔ You do not need to remember the names of polymers used as hydrogels, in shape memory polymers, or in waterproof clothing. But be prepared to answer questions about polymers and their properties from information given in the exam.

Key words

non-biodegradable, landfill site

▲ Waste plastic litters this beach

Did you know...?

It has been estimated that there are around 17 000 pieces of plastic per square kilometre of ocean, and that plastic bags take 1000 years to degrade.

▲ Landfill sites waste valuable land

Litter, litter, everywhere

Many polymers are unreactive. They do not react easily with acids, alkalis, or many other chemicals. They are often tough and strong. These properties are some of the reasons why polymers are so very useful. For example, we can store shampoo and water in plastic bottles without worrying that the bottle will dissolve. We can bury plastic gas pipes deep underground, knowing that they will serve us for many years. Unfortunately, these very benefits make polymers difficult to dispose of.

Decay bacteria help substances rot away. But many polymers are **non-biodegradable**. They cannot be decomposed by bacterial action. Plastic waste litters the Earth almost everywhere.

A What does non-biodegradable mean?

B Explain why many polymers are difficult to dispose of.

Disposing of polymers

When you throw out a used plastic item, the chances are it will end up in a **landfill site**. These are places where local authorities take waste to be buried. In the UK, around 85% of waste is dumped in landfill sites. They eventually become full of waste. It is difficult to find land for a new landfill site. Local people may object to it, for example because it will be unsightly. Although the plastic itself may not smell, other materials rotting away are smelly and may attract rats and gulls.

Crude oil is the raw material for most polymers. It is also the source of fuels such as petrol. Waste plastic can be burnt to release heat energy. This can be used to generate electricity or to heat buildings. The plastic must be burnt at a high temperature to stop toxic gases being made. Unless the heat energy released is used, disposal by burning is a waste of a valuable resource.

▲ Waste plastic can be burnt in incinerators like this

It is possible to recycle polymers. For example, PET is the tough plastic used to make drinks bottles. It can be recycled to make fibres for clothing. Unfortunately, plastic waste is usually a mixture of different polymers. It must be sorted into separate types of polymer, often by hand. This is difficult and expensive to do.

Chemists to the rescue

Chemists are developing ways to make addition polymers such as poly(ethene) biodegradable. Starch can be added to the polymer during manufacture. Bacteria can break down starch once the polymer gets wet. This causes the plastic item to break down into very small pieces. It has not rotted away, but it is no longer litter.

Bags are also now being made of cornstarch, a material obtained from maize. Cornstarch is biodegradable. However, some people think it is not ethical to grow maize to make bags when some people in the world do not have enough to eat.

PET	high-density poly(ethene)	PVC	low-density poly(ethene)	poly(propene)	polystyrene

PET	HDPE	PVC	LDPE	PP	PS

▲ Recycling symbols found on plastic items

Exam tip  **AQA**

✔ Make sure you can describe the drawbacks of different ways to dispose of waste polymers.

Did you know...?

It takes 25 PET bottles to make enough fibre for a fleece jacket.

Questions

1 Describe three ways in which waste polymers can be disposed of.

2 Explain why recycling symbols are put on plastic items.

3 Describe the benefits of biodegradable polymers.

4 Explain why burying waste plastics or burning them is a waste of valuable resources.

5 In 2002, Ireland introduced a tax on plastic carrier bags. Sales of carrier bags fell by 90%, but sales of other bags increased by 400%.

   (a) Suggest why the tax was introduced.

   (b) Suggest why plastic bag sales changed in the way they did.

Learning objectives

After studying this topic, you should be able to:

✔ evaluate the advantages and disadvantages of making ethanol from renewable and non-renewable sources

▲ This person has consumed too much ethanol

Key words

fermentation, hydration, glucose, natural catalyst

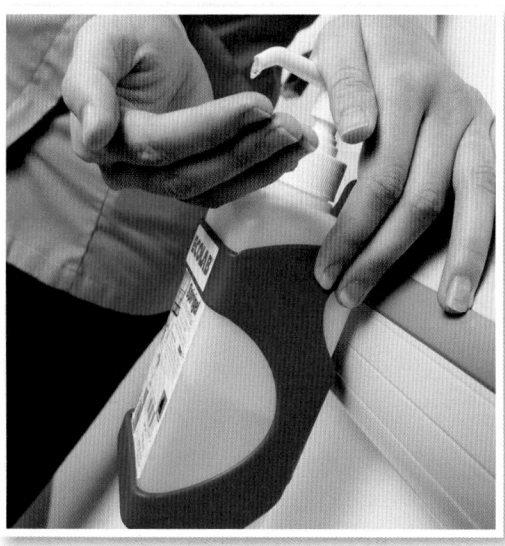

▲ Ethanol is an effective disinfectant

Using ethanol

Ethanol is big business. It makes huge profits for the companies that produce it. But its costs are huge, too. In 2005 ethanol was responsible for around 26 600 deaths in England and Wales. Ethanol costs the National Health Service an estimated £3 billion each year.

So what is this stuff called ethanol? Its formula is C_2H_5OH. It is the main ingredient in alcoholic drinks. It relaxes people and makes them lose their inhibitions.

Ethanol is not just used in drinks. It has many other uses, including:

• as a solvent in perfumes and after-shave
• as a disinfectant in hospital hand gels
• as a fuel for cars (see spread C1.16).

> **A** List four uses of ethanol.
>
> **B** Describe one problem associated with ethanol.

Making ethanol

There are two ways of making ethanol. People have been using one of these methods – **fermentation** – for many centuries. The other method was developed more recently.

Ethanol from ethene

The cracking of crude oil fractions produces huge amounts of ethene gas. Chemical companies react some of this ethene with steam to make ethanol. A catalyst, usually phosphoric acid, speeds up the reaction. The reaction works well at about 300 °C. These equations represent this **hydration** reaction:

$$\text{ethene} + \text{steam} \rightarrow \text{ethanol}$$
$$C_2H_4(g) + H_2O(g) \rightarrow C_2H_5OH(g)$$

The process is efficient because there are no waste products. It is a continuous process – it carries on for as long as ethene and steam are being added. However, ethene is a non-renewable resource. It is made from crude oil, which will run out one day.

Ethanol made from ethene is not normally used in alcoholic drinks. It can be used as a fuel and as a solvent.

Ethanol from sugars

Wine is made from grapes by fermentation. The grapes contain sugars, including **glucose**. Enzymes in yeast break down the glucose into ethanol and carbon dioxide. The ethanol mixes with other chemicals in the grapes to make wine. Fermentation works best at a temperature of 37 °C. The yeast enzymes are a **natural catalyst**. The raw materials for fermentation can be grown again, so they are renewable.

$$\text{glucose} \rightarrow \text{ethanol} + \text{carbon dioxide}$$
$$C_6H_{12}O_6(aq) \rightarrow 2C_2H_5OH(l) + 2CO_2(g)$$

Ethanol for fuel is often made by the fermentation of crops such as sugar cane.

> C List the temperatures and catalysts for the two methods of making ethanol.

Making ethanol – which way is better?

Which is the better way of making ethanol? Each method has its pros and cons.

The crude oil that makes ethene is non-renewable

It is not right to use land to grow plants for ethanol fuel. The land is better used for crops.

I work in a fermentation factory. Making ethanol by fermentation provides many more jobs than making ethanol from ethene.

Of the two processes, fermentation needs less energy because it happens at a lower temperature

▲ The ethanol in cider is made from sugars in apples

Did you know...?

Wasps can get drunk in autumn when they feed on fermented sugars from rotting apples.

Exam tip   AQA

✔ Remember the advantages and disadvantages of making ethanol by the two methods.

Questions

1 Name the raw materials for making ethanol from ethene. ↓ E

2 Name the waste product formed when ethanol is made by fermentation.

3 Draw up a table to compare the advantages and disadvantages of making ethanol by the two methods. ↓ C

4 Write a balanced symbol equation to summarise the two methods by which ethanol can be produced. ↓ A*

Learning objectives

After studying this topic, you should be able to:

✔ evaluate the impacts of using vegetable oils in foods

A List three ways of using plant oils in food and cooking.

B Explain why sunflower oil is suitable for frying.

Edible oils

What's the connection between the field of flowers below, and the bus and the plate of chips on the left?

The flowers produce rapeseed oil that fuels the bus and cooked the chips. Rapeseed oil is just one of many types of oil obtained from plants. In 2008, humans used more than 18 million tonnes of this oil.

Oil matters

Plant oils are important. We heat oils such as sunflower oil, rapeseed oil, and soya bean oil and use them to cook food. Oils are used for cooking because they have higher boiling points than water. So foods cook more quickly in oils. Oil-cooked foods have different flavours from those cooked in water – think about the difference in flavour between boiled potatoes and chips, for example.

Sunflower, rapeseed, and soya bean oils are particularly suitable for frying because they have high flash points. This means they catch fire only at high temperatures. You need to heat sunflower oil to 274 °C before it will set itself alight, for example.

People add olive oil to salads because they enjoy its flavour. Plant oils are vital ingredients in many types of pastries and biscuits, too. They help give these foods their crumbly texture.

Oil benefits

Oils are high in energy. If you eat lots of oil-rich food, you may put on weight. On the other hand, if you're planning an expedition to the North Pole, you may decide to take with you foods that are high in oil. Your body needs extra energy from food in cold climates or if you are physically active.

The table shows the energy content of sunflower oil compared to other foods.

Food	Energy in 100 g of the food (kJ, approximate values)
sunflower oil	3800
potato crisps	2100
apple	200
chicken breast	600

C List the foods in the table according to how much energy they provide.

D Suggest why potato crisps have a high energy content.

Plant oils provide **nutrients** as well as energy. Sunflower oil is high in vitamin E. There is evidence that vitamin E is an antioxidant. This means that it protects cells membranes from damage and may help prevent cancer.

Olive oil contains a compound called oleocanthal. Scientists have found that oleocanthal has anti-inflammatory properties. They suggest that the compound may explain why there is less heart disease in countries where people eat lots of olive oil.

Some plant oils are a good source of omega-3 fatty acids. Many scientists have researched the benefits of these substances. They found evidence that omega-3 fatty acids may help to prevent cancer, heart disease, and even poor behaviour.

Did you know...?

Scientists recently discovered that early artists used paints based on walnut oil and poppy seed oil to create the earliest known painting in a cave in Afghanistan. The painting dates from about 650 AD.

Key words

nutrient

Exam tip  AQA

✓ Remember – plant oils can come from fruits, seeds, and nuts. They provide energy and nutrients.

Questions

1 List three plants that provide oils.

2 Describe three health benefits of plant oils.

3 Explain the benefits of cooking foods in plant oils.

4 Describe one way in which plant oils may harm health.

5 Write a paragraph to evaluate the impacts of plant oils on health.

↓ E
↓ C
↓ A*

25: Oils from fruit and seeds

Learning objectives

After studying this topic, you should be able to:

- ✔ describe how plant oils are extracted from fruits, seeds, and nuts

Where do plant oils come from?

Plants store oils in their fruits, seeds, and nuts. The oil is part of a seed's food store. The seed needs the oil as an energy source when it starts to germinate.

▲ Olive oil is extracted from the fruit of olive trees

▲ Peanut oil is extracted from peanuts

How do we extract oils from fruits, seeds, and nuts? Read on to find out.

Extracting plant oils

Pressing

People have been extracting oil from olives for centuries. Roman olive presses still exist in Morocco. Today, traditional olive oil producers continue to use crushing and **pressing** to separate oil from the substances it is mixed with in olives. Here's how:

- Crush the olives into a paste to release oil from the cell vacuoles.
- Mix the paste for 30 minutes so that small oil droplets join together to form bigger ones.
- Press the paste by spreading it onto fibre discs and applying pressure to squash out the liquids.
- Collect the liquid mixture – of mainly oil and water – and allow it to settle. The oil floats on the water and is poured off.
- Remove impurities from the olive oil.

▲ Sunflower oil is extracted from sunflower seeds

▲ A traditional olive oil press

Solvent extraction and distillation

Extracting oil from sunflower seeds is less straightforward. The process has several stages:

- Remove the hulls from the seeds.
- Press the seeds to obtain some oil.
- Add a solvent, such as hexane, to the solid that remains. Any oil still mixed with the solid will dissolve.
- Use **distillation** to separate the solvent from the oil that is dissolved in it.

> A Explain how olive oil is separated from olives by crushing and pressing.
>
> B Explain how sunflower oil is extracted from sunflower seeds in a multi-step process.

Steam distillation

You can extract lavender oil by **steam distillation**. Here's how.

Set up the apparatus like this.

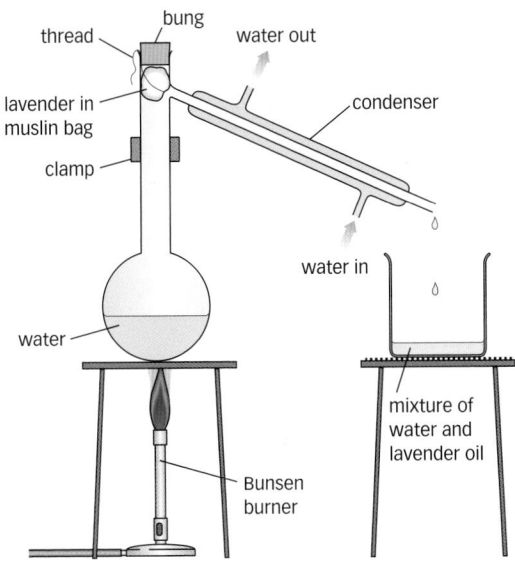

- Heat the water so that it boils and begins to make steam.
- Allow the steam to pass through the plant material. The volatile oils in the lavender vaporise.
- The steam and plant oil vapour move to the condenser. Here, they cool down and become liquid.
- The beaker contains lavender oil and water.

Key words

pressing, distillation, steam distillation

Exam tip AQA

✔ Remember – olive oil is extracted by pressing and sunflower oil is extracted by pressing, solvent extraction, and distillation. You do not need to learn the details of these processes off by heart.

Questions

1 Name one fruit, one nut, and one seed from which oils are extracted. ↓E

2 List the stages used to extract sunflower oil from sunflower seeds. ↓C

3 Identify the changes of state that happen in the steam distillation process. Write down the stages of the process at which each change of state occurs. ↓A*

Learning objectives

After studying this topic, you should be able to:

✔ explain how the properties of emulsions make them suitable for their uses

✔ explain how emulsifiers work

▲ What do these three substances have in common?

A Describe what happens when you mix oil and water.

B Explain why oil and water don't mix well.

Salad dressing, paint, and ice cream

What do salad dressing, ice cream, and emulsion paint have in common?

They are all emulsions. But what are emulsions, and what are they like inside? Read on to find out more.

Mixing oil and water

Vegetable oils and water don't mix together well. However hard you shake them together, the liquids separate out when you stop shaking. Oils are less dense than water, so they float on its surface.

Oil and water don't mix because their particles are too different. Vegetable oil molecules include long hydrocarbon chains. These chains cannot interact with small water molecules.

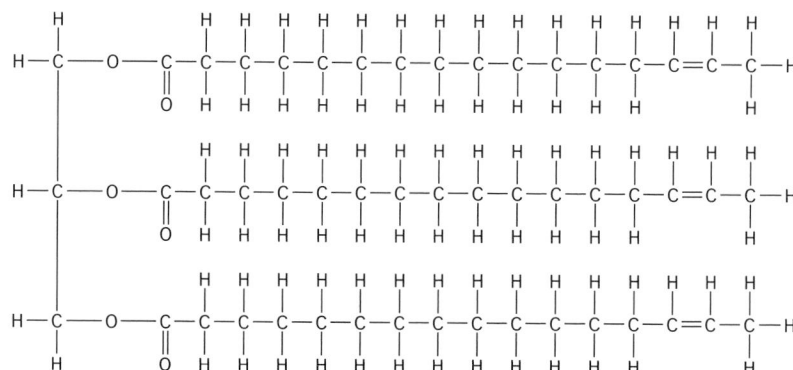

▲ A typical vegetable oil molecule

Inside emulsions

If you add an **emulsifier** to oil and water and shake well, the oil and water no longer separate out. The emulsifier stabilises the mixture and an **emulsion** forms.

Emulsions are more viscous, or thicker, than the liquids they are made from. This property leads to emulsions having a wide range of uses:

- Salad dressing contains vinegar, oil, and an emulsifier. Its high viscosity means it coats lettuce leaves well.
- Ice cream is an emulsion. The properties of the emulsion contribute to the special texture and appearance of ice cream.
- Emulsion paint coats walls well because of its particular texture and high viscosity.

- Many cosmetics are emulsions. Hand cream, moisturising cream, and shaving cream are oil-in-water emulsions. They consist of tiny oil droplets dispersed in water.
- More greasy cosmetics, such as sunscreens, are usually water-in-oil emulsions. They are made up of water droplets dispersed in oil.

Emulsifier safety

You've probably heard of some emulsifiers. Mayonnaise, for example, contains egg. Other emulsifiers are less well-known.
- Carrageenan (E407) is an emulsifier extracted from red seaweed. It is an ingredient of many processed desserts and ice cream.
- Tragacanth (E413) is a gum obtained from the sap of a plant that grows in Iran. It is used in cake decorations.

Artificial emulsifiers are identified by **E-numbers**. Additives with E-numbers have been safety tested and licensed by the European Union.

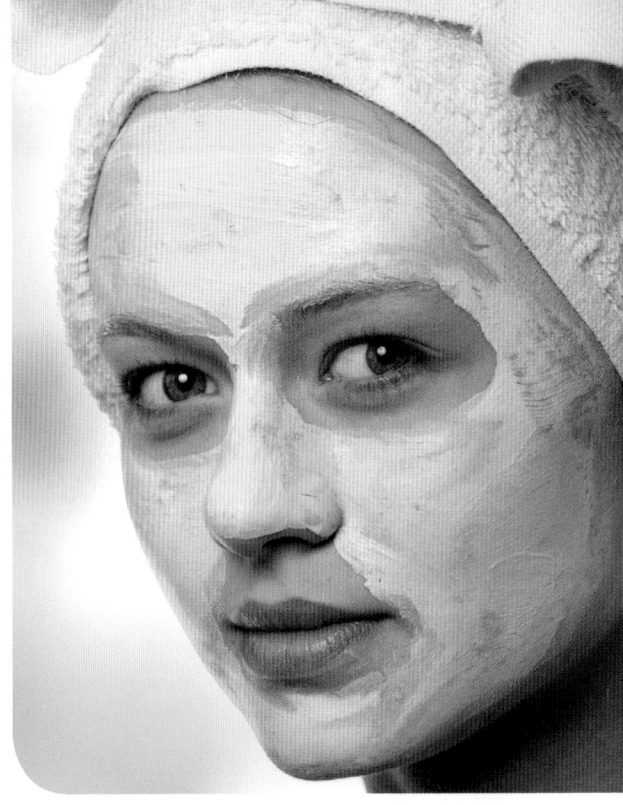
▲ Many cosmetics are emulsions

Key words

emulsifier, emulsion, E-number, **hydrophilic**, **hydrophobic**

Emulsifiers at work

Emulsifier molecules have two different ends:
- One end interacts well with water molecules. This is the **hydrophilic** end.
- The other end interacts well with oil molecules, and badly with water molecules. This is the **hydrophobic** end.

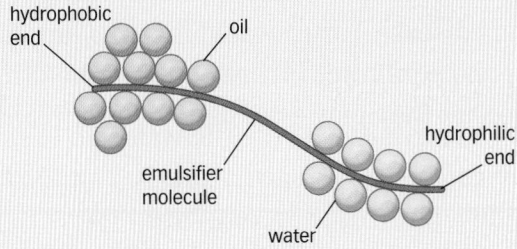

In an oil-in-water emulsion, emulsifier molecules coat the surface of oil droplets. The hydrophobic ends of the emulsifier molecules interact with the oil droplets. The hydrophilic ends interact with the water. These coatings keep the oil droplets evenly dispersed throughout the emulsion, and stop them clumping together to form their own separate layer.

Exam tip

✓ Remember that emulsifiers stop emulsions separating out.

Questions

1 Describe the property of emulsions that makes them suitable for salad dressings.

2 Explain why plants make and store oils.

3 Explain how emulsifiers stop oil and water separating out in emulsions.

Learning objectives

After studying this topic, you should be able to:

✔ explain the differences between saturated fats and unsaturated fats

✔ describe how to make margarine from vegetable oil

▲ Bread with olive oil …

▲ … and toast with butter

Spread for your bread

Riana likes butter on her toast. Matthew prefers margarine. Donnatella drizzles olive oil over bread for a delicious snack. Butter, margarine, and olive oil provide similar amounts of energy. They are all mixtures of compounds consisting of atoms of carbon, hydrogen, and oxygen. What makes them different?

Liquid or solid?

Butter and margarine are solid at room temperature. Most plant oils are liquid at room temperature. They melt at lower temperatures than butter and margarine.

The structures of the molecules in fats and oils explain their different melting temperatures.

- Butter is a **saturated** fat. There are no double bonds in its molecules.
- Most plant oils have **unsaturated** hydrocarbon chains in their molecules. The hydrocarbon chains contain double carbon–carbon bonds.
  1. **Monounsaturated** fats have one double bond per hydrocarbon chain.
  2. **Polyunsaturated** fats have several double bonds per hydrocarbon chain.

Good for your health?

Saturated fats raise blood cholesterol and so increase the risk of heart disease. Unsaturated fats in plant oils such as olive oil and sunflower oil are better for health. There is evidence that they may even help to lower blood cholesterol.

> A Explain the difference between saturated fats and unsaturated fats.
>
> B Explain why sunflower oil is better for your health than butter.

Detecting double bonds

Achita has three plant oils. She needs to know if any of them are high in saturated fats.

Achita adds orange bromine water to samples of the oil. The test is the same as that for detecting double bonds in alkenes (see spread C1.18). Here are Achita's results.

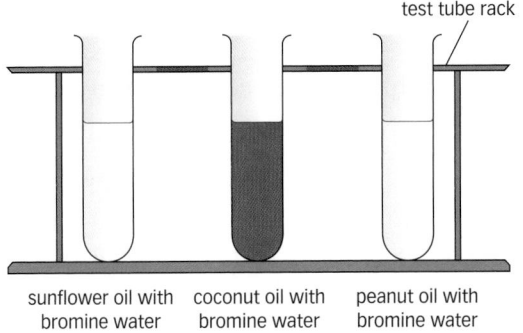

test tube rack

sunflower oil with   coconut oil with   peanut oil with
bromine water   bromine water   bromine water

The bromine has reacted with sunflower oil to form a colourless mixture. This shows that sunflower oil is unsaturated. Bromine atoms have added to the carbon atoms on either side of the double bonds.

There is no change in the coconut oil test tube. Bromine has not reacted with the oil. So coconut oil is saturated. It contains no carbon–carbon double bonds.

Making margarine

Food companies harden unsaturated vegetable oils by adding hydrogen gas to them. These reactions happen at about 60 °C. A nickel catalyst speeds up the reaction.

In these **hydrogenation** reactions, hydrogen atoms add to carbon atoms on both sides of the double bonds.

▲ The diagram shows just part of a plant oil molecule

Hydrogenation reactions convert unsaturated oils into saturated ones. The saturated oils have higher melting points. They are solid at room temperature. They can now be used as spreads and to make cakes and pastries.

C Explain what is shown by the result of the test of bromine water with peanut oil.

Exam tip

✓ Molecules in saturated oils have no double bonds. Unsaturated oil molecules do have double bonds. Bromine water tests for unsaturation.

Questions

1   What is a saturated fat?

2   Describe how to test an oil for unsaturation. Explain what different results would mean.

3   Describe the conditions needed for hydrogenation reactions.

4   Predict whether butter or olive oil is higher in unsaturated fats. Explain your decision.

28: Inside the Earth

Learning objectives

After studying this topic, you should be able to:

✔ describe the structure of the Earth

✔ describe Wegener's theory of crustal movement (continental drift) and explain why other scientists did not at first accept it

The structure of the Earth

Think about all the things you've used today. Where did they all come from? The answer is the Earth, its **atmosphere**, and the oceans. All the minerals and other resources humans use come from one of these three sources.

The Earth is a sphere. It is made up of several layers. The main ones are:

- a rocky **crust**, which is thin compared to the other layers
- the **mantle**, which goes down almost halfway to the centre of the Earth; the mantle has solid properties, but can flow very slowly
- the **core**, which is made of iron and nickel.

Surrounding the Earth is a mixture of gases. This is the atmosphere.

The radius of the Earth is about 6370 km. Imagine you could ride in a jumbo jet to the centre of the Earth. The journey would take about seven hours at top speed, and there would be just enough fuel to return to the surface. But it is not really possible to study the Earth's structure directly. The deepest mine is only 3.5 km deep.

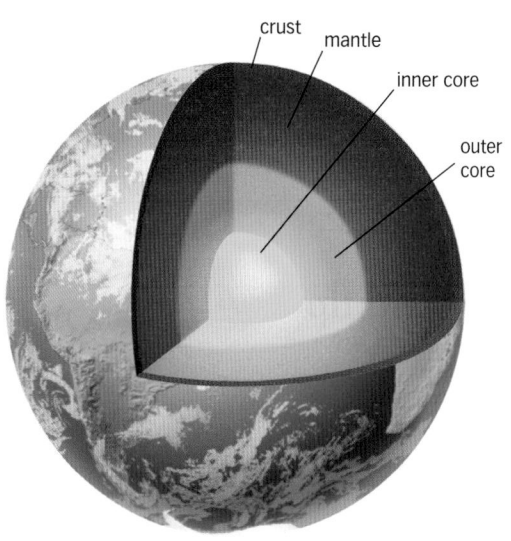

crust
mantle
inner core
outer core

▲ The structure of the Earth

A Describe the structure of the Earth, starting at its centre.

B Imagine the Earth as an egg. Which parts of the Earth can be represented by the egg shell, the white, and the yolk?

C Explain why it is not possible to study the Earth's structure directly.

▲ A model for the Earth?

A shrinking Earth?

People have been wondering about the Earth for centuries. Where did mountains come from? Why are continents the shapes they are?

Scientists once believed that the features of the Earth's surface were the result of the shrinking of the crust as the Earth cooled down after it was formed. Valleys and mountains were like the wrinkles on the surface of a drying apple. However, scientists have since collected evidence to show that this theory must be false.

Controversial theory

In 1912, the German scientist Alfred Wegener put forward a new theory to explain the history of the Earth. He suggested that the continents were once joined together, and that they had gradually moved apart. Wegener supported his theory with evidence:

- The shapes of Africa and South America look as if they might once have fitted together.
- Fossils of the same plants had been found in both Africa and South America.
- The two continents have the same rock types at the edges where they might have been joined.

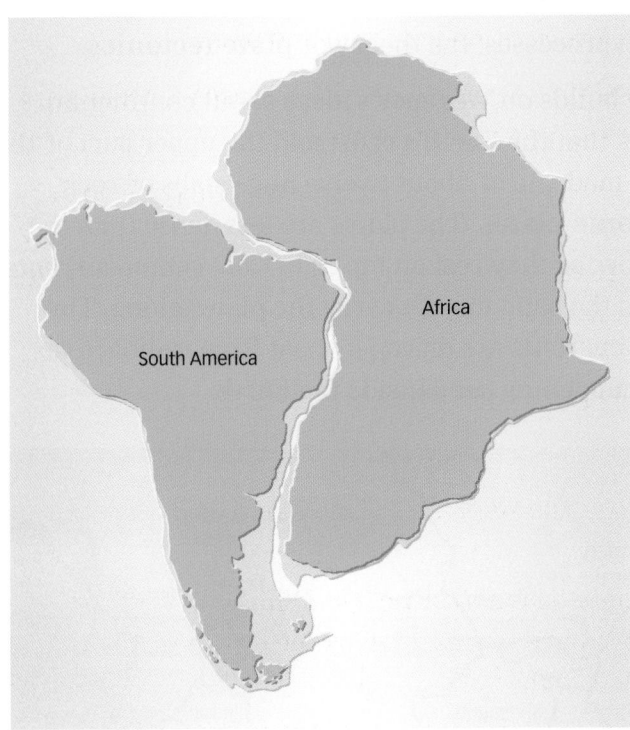

South America

Africa

At the time, most other scientists did not accept Wegener's theory of **crustal movement**, or **continental drift**. They could not see how the continents might have moved. They were also reluctant to believe Wegener because he was not a geologist.

However, by the 1950s, scientists were beginning to support Wegener's theory. They discovered evidence suggesting that the Atlantic Ocean sea floor was spreading. They discovered convection currents in the Earth's mantle. And they discovered similar rocks and fossils between different pairs of continents.

Did you know...?

The deepest hole ever drilled is the Kola Superdeep Borehole in Russia. It reached 12.261 km through the crust in 1989. By then it was too hot for drilling to continue.

Key words

atmosphere, crust, mantle, core, crustal movement, continental drift

Questions

1 List two parts of the Earth's structure which are solid.
2 Describe the Earth's mantle.
3 Describe Wegener's theory of continental drift.
4 List three pieces of evidence that support Wegener's theory of continental drift.
5 Explain why scientists were reluctant to accept Wegener's theory at first.

Learning objectives

After studying this topic, you should be able to:

- describe the theory of continental drift
- explain how moving tectonic plates cause volcanoes and earthquakes and make mountains
- explain why scientists cannot accurately predict when earthquakes and volcanic eruptions will happen

▲ Haiti disaster

Key words

earthquake, volcano, plate tectonics, tectonic plate, focus

Restless Earth

12 January 2010: A catastrophic magnitude 7.0 **earthquake** strikes Haiti. By 24 January, at least 52 aftershocks had been recorded. A quarter of a million people die and the same number are injured. A million more lose their homes.

23 April, 1902: Mount Pelée **volcano** in Martinique begins to erupt. More than 29 000 people die in its ash flows.

What caused these disasters? Could they have been predicted?

Tectonic plates

One big idea explains most volcanoes, earthquakes, and other Earth processes: the theory of **plate tectonics**.

The theory builds on Wegener's ideas about continental drift. It says that the Earth's crust and the upper part of the mantle are made up of about twelve huge slabs of rock called **tectonic plates**. The plates are less dense than the mantle below, so they rest on top of it. Convection currents deep within the solid mantle carry the plates along. The convection currents are driven by heat from radioactive processes happening deep inside the Earth.

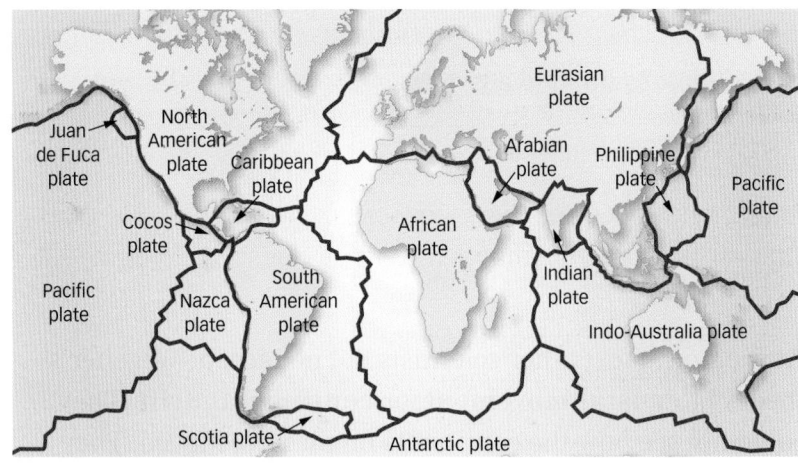

▲ Tectonic plates

Tectonic plates move very slowly, at a speed of a few centimetres a year. Global positioning satellites track their movements.

> A What is a tectonic plate?
>
> B Describe how tectonic plates are carried along.

Earthquakes

Earthquakes happen when tectonic plates move against each other suddenly.

At the San Andreas Fault in California, USA, two enormous tectonic plates are moving past each other in opposite directions. But, because of friction, the plates cannot slide smoothly and they sometimes get stuck. Huge forces build up as they keep trying to pass each other. Eventually the two plates overcome the frictional forces. They slip suddenly. There is an earthquake.

The place where the plates slip is the **focus** of an earthquake. Shock waves spread from the focus in all directions. The waves make buildings collapse. In earthquakes, falling buildings may kill and injure thousands of people.

Earthquakes are common at all moving plate boundaries. Undersea earthquakes may cause enormous waves – or tsunamis – which do great damage when they reach land.

Scientists cannot predict exactly when plates will suddenly slip. So people living on plate boundaries can expect an earthquake at any time.

Volcanoes

A volcano is a vent in the Earth's crust from which magma (liquid rock), ash, and gases such as carbon dioxide and sulfur dioxide erupt. Every year there are around 50 volcanic eruptions.

Scientists monitor active volcanoes closely to look for signs that an eruption may happen soon. A volcano may be about to erupt if
- there are earthquakes nearby
- the shape of the volcano changes
- the volcano gives off more gases than usual.

But scientists can still never be sure exactly when a volcano will erupt, or how severe eruptions will be.

Making mountains

Mountains may form when tectonic plates collide. For example, when two plates push against each other, rocks at the edge of one of the plates buckle and fold to form a mountain chain.

▲ San Andreas Fault

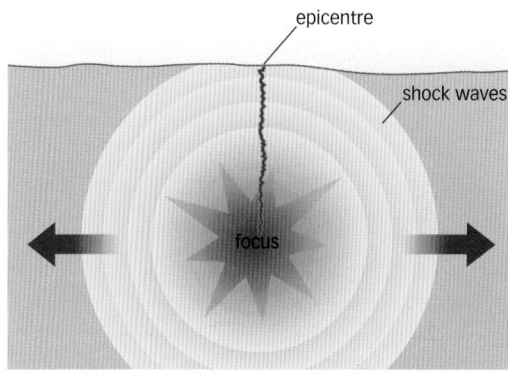
▲ Focus, epicentre, and shock waves of an earthquake

C Explain why scientists cannot predict exactly when an earthquake will happen.

D Describe the signs scientists look for to predict volcanic eruptions.

Questions

1 List two types of Earth events that can be explained by plate tectonics.

2 Describe how plates sliding past each other may cause earthquakes.

3 Suggest two actions a local council might take if scientists predict a volcanic eruption.

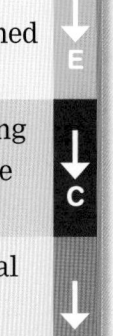

Learning objectives

After studying this topic, you should be able to:

✔ describe the composition of the Earth's atmosphere

✔ know how to separate gases from the air

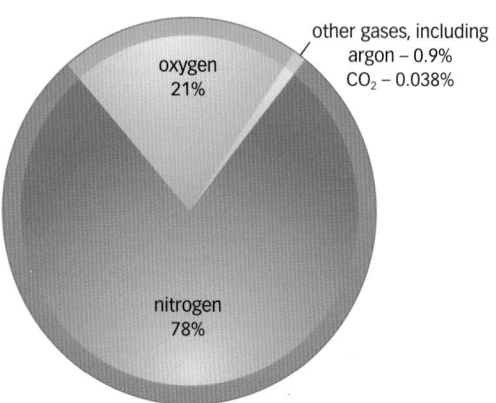

▲ The pie chart shows the proportions of the main gases in the air

A Make a table to show the proportions of the main gases in the air.

Exam tip

✔ For Higher Tier, you do not need to memorise the boiling points of the gases in the air. Just remember that their different boiling points mean they can be separated by fractional distillation.

Breathing on the Moon

July 1969. Humans land on the Moon. Astronauts Neil Armstrong and Buzz Aldrin wear breathing apparatus for the first ever Moon walks. Why?

Unlike the Earth, no gases surround the Moon. The Moon has no atmosphere.

The atmosphere today

The Earth's atmosphere is a mixture of gases surrounding the planet. The Earth's gravity stops these gases escaping into space. The lower part of the atmosphere is the air that we breathe. It has stayed much the same as it is today for the last 200 million years.

Just two gases, nitrogen and oxygen, make up about 99% of the air. They are both elements. There are smaller proportions of other gases in the air, including carbon dioxide, water, and noble gases such as argon.

Gases from the atmosphere

The atmosphere is a vital source of raw materials for many processes. Nitrogen makes ammonia for fertilisers. Hospitals use oxygen to treat patients. The noble gases have a great variety of uses. But how do companies separate the gases of the air?

Nitrogen, oxygen, and the other gases of the air have different boiling points.

Gas	Boiling point (°C)
nitrogen	−196
oxygen	−183
argon	−186

B Make a bar chart to show the boiling points of the gases in the table.

Their different boiling points mean that the gases can be separated. Here's how.

- Cool the air to −200°C in stages, so that its gases condense to form a mixture of liquids. As the air cools:
  - water vapour condenses and is removed
  - carbon dioxide freezes at −79°C and is removed
  - oxygen and nitrogen condense at their boiling points.
- Separate the mixture of liquid oxygen, liquid nitrogen, and liquid argon by fractional distillation.

Did you know...?

Foods such as potato crisps are packaged in nitrogen gas to increase their shelf life.

▲ Liquid nitrogen is used to freeze food

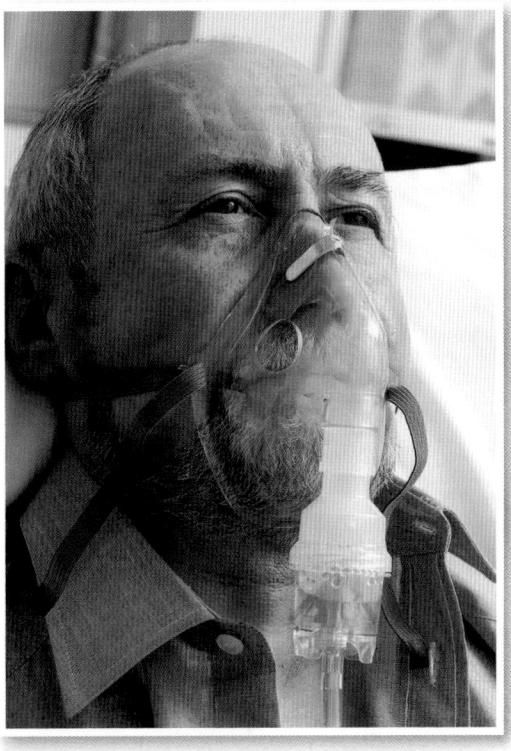

▲ Hospitals use oxygen to treat patients

▲ Filament light bulbs are filled with argon gas. They are being phased out because they are not energy efficient.

Questions

1. List the gases in the atmosphere and give their proportions.  E

2. Name three elements and two compounds that are present in the atmosphere. C

3. Describe how nitrogen and oxygen are obtained from the atmosphere.

4. Give one use each for nitrogen, oxygen, and argon. A*

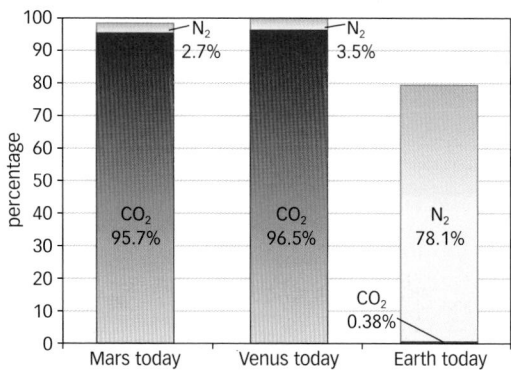

The Earth's early atmosphere was probably like the present day atmospheres of Mars and Venus

A Describe the ways in which the Earth's early atmosphere was like the present day atmospheres of Mars and Venus.

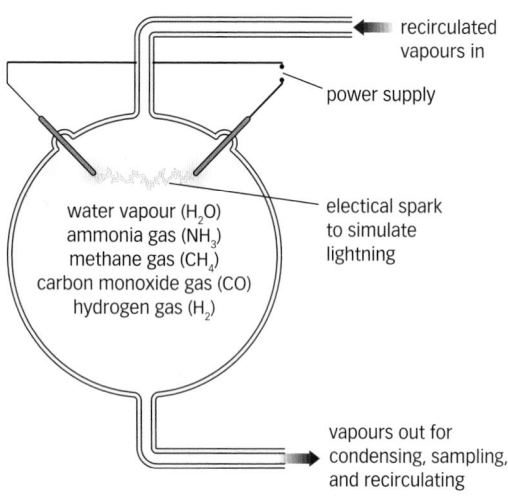

Part of the apparatus for the Miller–Urey experiment

Different theories

The Earth's atmosphere hasn't always been as it is today. So where has our atmosphere come from? How was it formed? Scientists have several theories to explain the origins of the atmosphere. We cannot know which, if any, is correct. No one was around to record events as they happened. One possible theory involves plants and volcanoes.

The first billion years

The Earth is about 4.5 billion years old. There was a lot of **volcanic activity** during its first billion years. When volcanoes erupt, they release huge quantities of gases. These are mainly water vapour and carbon dioxide, with smaller amounts of other gases such as ammonia (NH_3) and methane (CH_4).

It is likely that the Earth's early atmosphere came from gases released by volcanoes from inside the Earth's crust. The water vapour condensed to form the oceans. So the early atmosphere was mainly carbon dioxide with small proportions of ammonia and methane. There would have been little or no oxygen.

B Outline the main differences between the Earth's early atmosphere and its atmosphere today.

Life begins

Billions of years ago, life began. No one knows exactly how, but many scientists have collected evidence and suggested theories.

One theory, the **primordial soup theory**, suggests that gases in the early atmosphere reacted with each other in the presence of sunlight or lightning to make complex molecules that are the basis of life.

In 1953, two scientists designed an experiment to test the theory. The **Miller–Urey experiment** simulated a lightning spark in a mixture of the gases of the early atmosphere. A week later, more than 2% of the carbon in the system had formed compounds from which proteins in living cells are made. This, said the scientists, supports the primordial soup theory.

A changing atmosphere

Plants make their own food by **photosynthesis**. They take carbon dioxide from the atmosphere and release oxygen gas:

carbon dioxide + water → oxygen + glucose

As plants evolved from early living organisms, their photosynthesis reduced the amount of carbon dioxide in the atmosphere. Photosynthesis also increased the proportion of oxygen in the atmosphere until it reached today's level.

Photosynthesis was not the only reason why carbon dioxide levels decreased.

Locking up carbon in rocks

Carbon dioxide is a soluble gas, and large amounts of it dissolved in the oceans.

Shellfish and other sea creatures used some of this carbon dioxide to make their shells and skeletons. When the animals died, they fell to the bottom of the ocean. After many years, limestone, a **sedimentary rock**, formed from their shells and skeletons. The carbon atoms were locked away in limestone, mainly as calcium carbonate.

Locking up carbon in fossil fuels

Millions of years ago, dead plants and animals decayed under swamps. The dead organisms formed fossil fuels. The carbon atoms of the plants and animals were locked up in underground stores of coal, oil, and gas.

▲ Large amounts of carbon dioxide dissolve in the oceans

Key words

volcanic activity, **primordial soup theory, Miller–Urey experiment**, photosynthesis, sedimentary rock

▲ Limestone rock was formed from the shells and skeletons of shellfish and other sea creatures

Questions

1 Name four gases that were probably present in the Earth's early atmosphere. Where did these gases come from? ↓ E

2 Explain three ways by which carbon dioxide was removed from the Earth's early atmosphere.

3 Explain the origin of atmospheric oxygen. ↓ C

4 Explain why we cannot be sure how the Earth's atmosphere was formed.

5 Describe the primordial soup theory of the origin of life, and the experimental evidence that supports it. ↓ A*

Did you know...?

At current rates of photosynthesis it would take living things just 2000 years to make all the oxygen found in the Earth's atmosphere today.

Exam tip   AQA

✓ Remember that oxygen increased over time. Carbon dioxide decreased.

Learning objectives

After studying this topic, you should be able to:

✔ explain and evaluate the effects of human activities on the atmosphere

Key words

respiration, combustion, reservoir, deforestation

▲ Plants remove carbon dioxide from the atmosphere

Exam tip **AQA**

✔ Photosynthesis and dissolving in the oceans remove carbon dioxide from the atmosphere. Respiration and combustion add to atmospheric carbon dioxide.

Carbon dioxide – on the up

Carbon dioxide is vital to life. Without it, plants could not make their own food. Animals – including humans – would have nothing to eat. And Earth would be too cold for life as we know it.

Since about 1800, the percentage of carbon dioxide in the atmosphere has been increasing year by year. Why? And why does it matter?

Into and out of the atmosphere

Into the atmosphere

Two main processes add carbon dioxide to the atmosphere:

- **Respiration** is the process by which plants and animals release energy from food. Carbon dioxide is one of the waste products of this process.

    glucose + oxygen → carbon dioxide + water

- **Combustion** also releases carbon dioxide into the atmosphere. Fossil fuels contain carbon atoms that have been locked away for millions of years. When they are burnt, they join with oxygen atoms and enter the atmosphere as carbon dioxide molecules. For example:

    methane + oxygen → carbon dioxide + water

Leaving the atmosphere

Today, about 0.038% of the Earth's atmosphere is carbon dioxide. Plants remove some of this gas to make their own food by photosynthesis. Carbon dioxide is removed from the atmosphere when it dissolves in the oceans, too.

The carbon cycle

The carbon cycle summarises the processes that add and remove carbon dioxide from the atmosphere.

The carbon cycle also shows the stores – or **reservoirs** – of carbon, including the atmosphere, the oceans, sedimentary rocks, and fossil fuels.

> A Name two processes that add carbon dioxide to the atmosphere.
>
> B Name two processes that remove carbon dioxide from the atmosphere.

Off balance

You have just seen that some processes add carbon dioxide to the atmosphere and some remove it. If these processes balance, then the concentration of carbon dioxide in the atmosphere will not change.

Since 1800, these processes have not been balanced. Human activities have been adding carbon dioxide to the atmosphere faster than it is removed:

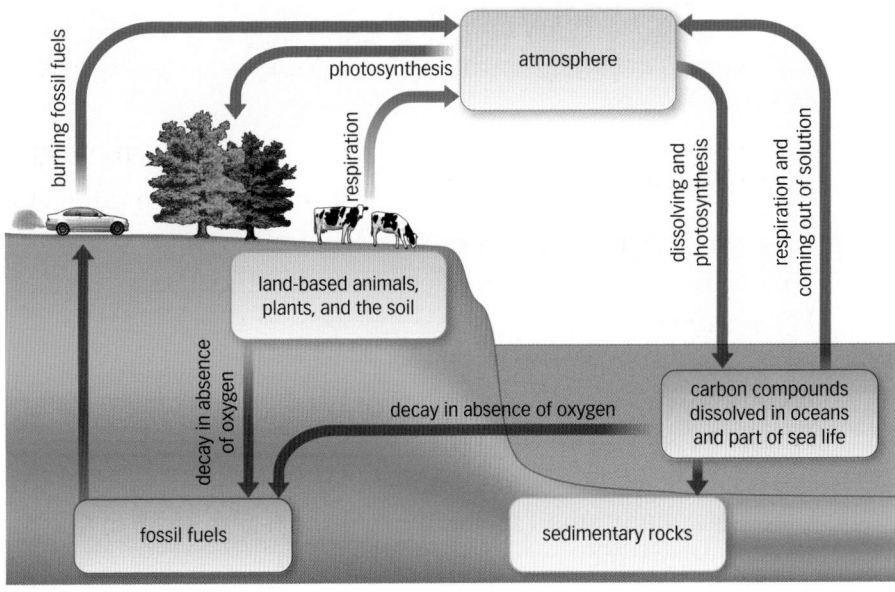

▲ The carbon cycle

- People burn fossil fuels to generate electricity, heat houses, and fuel vehicles.
- The human population is increasing. This adds to the demand for energy and leads to more fossil fuels being burnt. It also adds to the demand for land for buildings, roads, and farms.
- Forests are cut down, or burnt, to make way for human use. This **deforestation** has a big impact on carbon dioxide levels. When trees burn, they release carbon dioxide to the atmosphere. Fewer trees remain to remove carbon dioxide from the atmosphere by photosynthesis.

As the amount of carbon dioxide in the atmosphere has increased, so more carbon dioxide has dissolved in the oceans. This extra carbon dioxide makes seawater more acidic. This causes problems for some living organisms. Shellfish, for example, have difficulty making their shells.

Global warming

Most scientists agree that the increasing concentration of carbon dioxide in our atmosphere causes global warming. An increase in the Earth's average air temperature will have far-reaching consequences:
- Weather patterns will change. There are likely to be more extreme weather events. Some areas will suffer from drought, whilst others will flood.
- Polar ice caps will melt. Sea levels will rise, and low-lying coastal areas will flood.

▲ Deforestation, contributing to increased atmospheric carbon dioxide levels

Questions

1 Name a natural process that adds carbon dioxide to the atmosphere.

2 Name a human activity that adds carbon dioxide to the atmosphere.

↓ E

3 Make a table to summarise the ways by which carbon dioxide enters and leaves the atmosphere.

↓ C

4 Describe and explain the problems caused by more carbon dioxide dissolving in the sea.

↓ A*

Course catch-up

Revision checklist

- Cracking using heat and a catalyst breaks down long-chain alkanes into smaller molecules, including alkenes.

- Alkenes are unsaturated hydrocarbons which contain C=C double bonds and have general formula C_nH_{2n}. Bromine water is used to test for C=C bonds.

- Alkenes are used to manufacture polymers. In polymerisation reactions, many small monomer molecules join to form one large polymer molecule.

- Polymers may have a wide range of properties. The uses of a polymer depend on its properties.

- The disposal of non-biodegradable polymers causes environmental problems.

- Ethanol is a molecule which can be formed from renewable sources (fermentation of sugars) or non-renewable sources (reaction of ethene + steam).

- Plant oils have high energy content, high boiling point, and are used as foods and in cooking.

- Plant oils can be hardened into margarine for use in foods by reacting with hydrogen (hydrogenation).

- Plant oils can be extracted from plant material by pressing, solvent extraction followed by distillation, or steam distillation.

- Plant oils do not dissolve in water, but can be made into emulsions by using emulsifiers.

- The Earth has a layered structure. The thin rocky crust floats on the mantle, and at the centre is the hot iron core.

- The crust is divided into tectonic plates. Convection currents in the mantle cause the plates to move. Earthquakes and volcanic eruptions occur at plate boundaries.

- The Earth's atmosphere has a stable composition (78% nitrogen, 21% oxygen + small amounts of other gases, including CO_2).

- The early atmosphere of the Earth was mostly CO_2 and water vapour plus small amounts of ammonia and methane.

- Processes which caused the atmosphere to change include rock formation, development of plant life (photosynthesis), dissolving of gases in the seas, and volcanic eruptions.

- The formation of complex molecules from the reaction of hydrocarbons with ammonia during lightning strikes may have helped life to form.

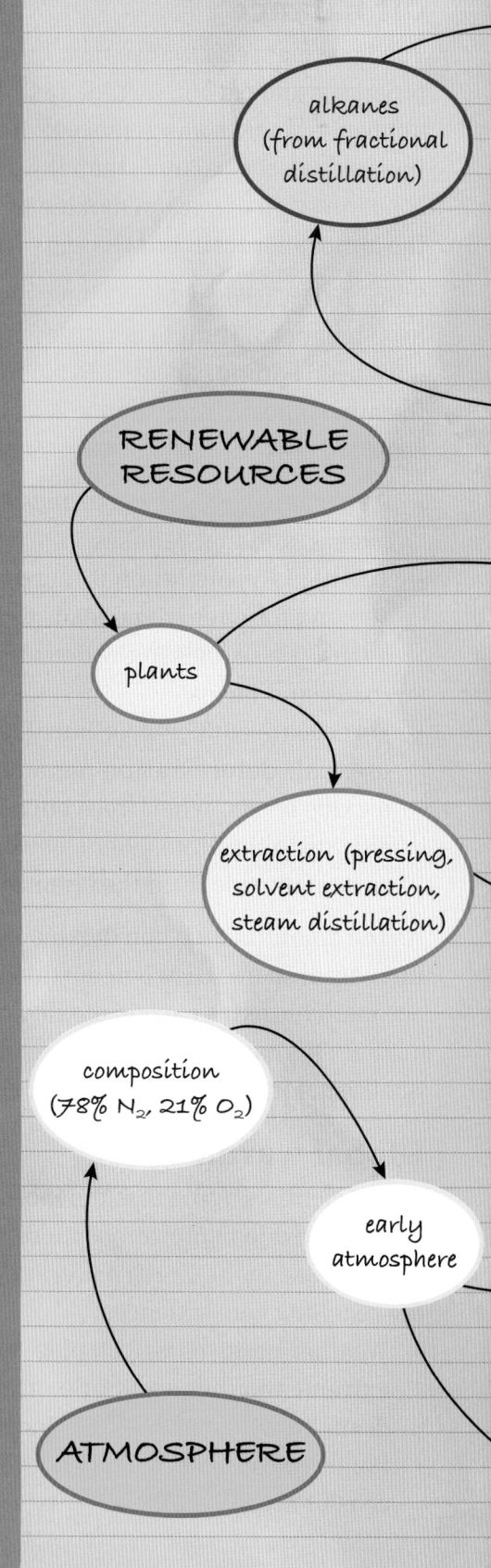

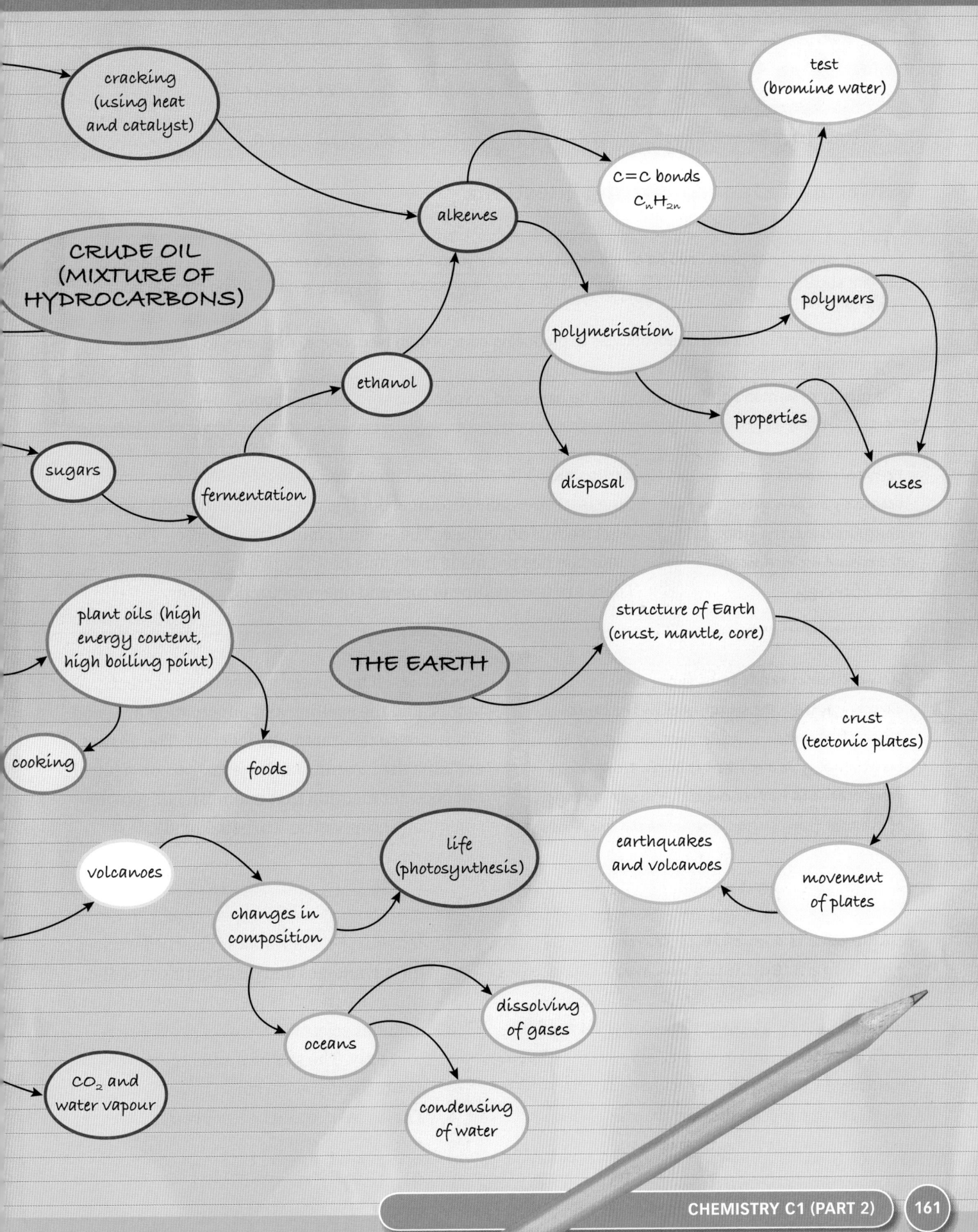

cracking (using heat and catalyst)

CRUDE OIL (MIXTURE OF HYDROCARBONS)

alkenes

$C=C$ bonds C_nH_{2n}

test (bromine water)

polymerisation

polymers

properties

disposal

uses

ethanol

sugars

fermentation

plant oils (high energy content, high boiling point)

cooking

foods

THE EARTH

structure of Earth (crust, mantle, core)

crust (tectonic plates)

volcanoes

changes in composition

life (photosynthesis)

earthquakes and volcanoes

movement of plates

CO₂ and water vapour

oceans

dissolving of gases

condensing of water

AQA Upgrade

Answering Extended Writing questions

In Brazil, ethanol has been used as a vehicle fuel for many years. Ethanol can be produced by the fermentation of plant sugars, such as those in sugar cane. It can also be produced from ethene gas. Ethene is made from crude oil.

Outline the advantages and disadvantages of producing ethanol from plant crops, compared to producing ethanol from ethene.

The quality of written communication will be assessed in your answer to this question.

G–E

Fermentasion is good because you can grow shugar cane every year. The ethene method needs lots of energy

Examiner: The candidate knows that ethanol made from sugar cane is a renewable resource, but has not used the scientific word to describe this advantage. The second point is also correct. The candidate has not made it obvious how the two processes compare. There are two spelling errors and one punctuation error.

D–C

Using ethanol made from sugar cane is carbon neutral. The sugar take in the same amount of carbon dioxide when they grow as the ethanol give out when it burn. Ethanol from ethene is not renewable. A disadvantage of ethanol from sugar is you should use land to grow food.

Examiner: The answer makes three correct points, and the spelling and punctuation are good. There are some grammatical errors. The answer explains the meaning of the term 'carbon neutral', but does not mention that making fertilisers for sugar cane crops causes carbon dioxide emissions.

B–A*

Sugar cane is a renewable resource – you can grow it again. Ethene is a non-renewable resource, because it comes from crude oil. Fermentation happens at 37ºC. Ethene makes ethanol by reacting with steam at 300ºC. So fermentation needs lower energy inputs. But fermentation makes waste carbon dioxide. The other process makes no waste products. It is morally wrong to use land for fuel crops instead of food.

Examiner: This answer clearly describes the advantages and disadvantages of the two processes. It is logically organised, and includes scientific terms that are used correctly. The spelling, punctuation, and grammar are accurate.
The answer would be even better if it made clear which statements refer to advantages and which to disadvantages.

Exam-style questions

1 Scientists now know that the Earth is made up of several layers:

core   crust   mantle

A01 **a** Use the words above to label the different layers in this diagram.

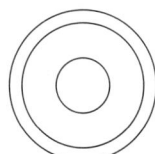

A01 **b** Which layer is:
  i made of nickel and iron?
  ii a source of important raw materials for the chemical industry?
  iii made up of a number of large tectonic plates?

2 Large hydrocarbon molecules obtained from crude oil can be converted into smaller molecules by a process called cracking.

A01 **a** What conditions are needed for a cracking reaction?

A02 **b** Give one reason why cracking is important in the petrochemical industry.

3 **a** The displayed formula of chloroethene is:

$$H_2C=CHCl$$

A02 **i** Why is this molecule described as unsaturated?

A02 **ii** A student adds bromine water to a small sample of this substance. What would she observe?

iii Complete this balanced equation to show the reaction which happens when the bromine is added:
$$C_2H_3Cl + Br_2 \rightarrow$$

b Chloroethene is a monomer used in industry to manufacture the polymer poly(chloroethene).

A02 **i** Draw out the displayed formula of poly(chloroethene).

A02 **ii** Poly(chloroethene) is used to make window frames. Suggest two properties which the polymer must have to make it suitable for this use.

Extended Writing

4 Olives grow in some parts of the world. They are a good source of oil. This olive oil is used in cooking, is eaten as a food, and can be used as a fuel.
A01 Write about why olives can be used in this way.

5 Polymers are a very important type of substance in today's society. However, many people are worried about how to dispose of them.
A01 Explain why disposing of polymers is a particular problem.

6 The atmosphere of the Earth today contains about 21% oxygen, 78% nitrogen, and 1% other gases.
A02 Describe how the early atmosphere may have been different to today's atmosphere.

A01 Recall the science

A02 Apply your knowledge

A03 Evaluate and analyse the evidence

G–E

D–C

B–A*

B–A*

D–C

G–E

D–C

B–A*

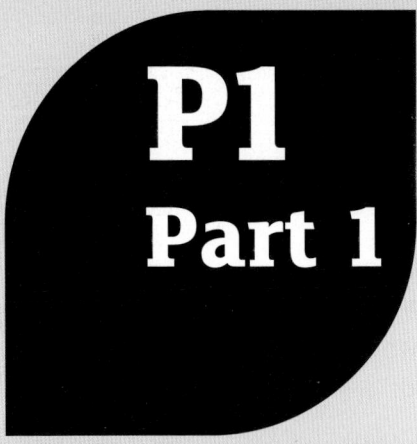

P1
Part 1

Energy and efficiency

Why study this unit?

Energy can be transferred in different ways by heating processes. The clothes that you wear on a cold day keep you warm by reducing the transfer of heat energy from your body. Houses are insulated to reduce the amount of heat transferred to the outside. But some appliances need to transfer lots of energy – central heating radiators transfer energy into a room.

In this unit you will look at the different ways that energy can be transferred by conduction, convection, radiation, evaporation, and condensation. You will learn how these transfers are minimised in homes and some of the things you use, but maximised in other appliances. You will also learn about the useful and wasted energy transferred by appliances and how to calculate the efficiency of an appliance.

Finally, you will learn about how different appliances in your home transfer electrical energy into a range of different forms of energy, and how much these appliances cost to run.

You should remember

1 Energy cannot be created or destroyed.

2 Energy can be transferred by a number of different methods.

3 The hotter an object, the higher its temperature.

4 Heating or cooling involves a transfer of energy.

5 There are three states of matter: solids, liquids, and gases.

This house in Kent is so well insulated that it does not need any central heating. The house uses energy from the Sun and stores it for when the weather is cold. The house uses less than 10% of the energy that a conventional three-bedroom house uses for heating rooms and water. Most of the energy used by this house is for heating hot water. The house has a ventilation system which takes in fresh air from outside and transfers heat energy to it from the stale air that is about to be pumped outside.

Learning objectives

After studying this topic, you should be able to:

- ✔ understand that all objects both emit and absorb infrared radiation
- ✔ investigate which surfaces are good or bad at absorbing infrared radiation

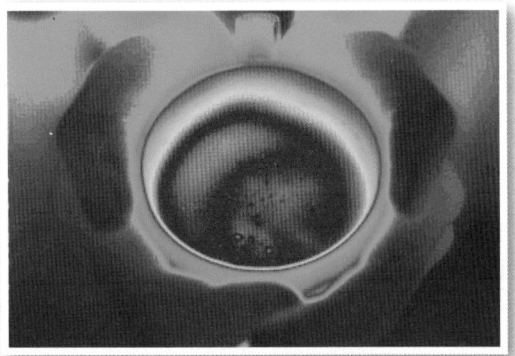

▲ A thermogram of a hot drink

Key words

infrared radiation, thermogram, emit, medium

▲ This cooker ring is emitting infrared radiation and light

Infrared radiation

When you put your hands on a cup that contains a hot drink, heat energy is transferred to your hands by conduction. If you take your hands off the cup, but keep them close to it, you can still feel heat energy being transferred from the cup.

This energy is being transferred by **infrared radiation** (IR). Infrared radiation is a type of electromagnetic wave.

A **thermogram** shows how much infrared radiation something is emitting. You take a picture using a special camera which is sensitive to infrared radiation but not visible light. The photo on the left shows a thermogram of a hot drink. The hotter areas are red. The cooler areas are blue.

> **A** What is infrared radiation?

All objects **emit** and absorb infrared radiation.

When something is glowing then it is also emitting light radiation. The electric cooker ring in the photo is glowing red – it is emitting light as well as infrared radiation.

Sometimes you feel infrared radiation but no light radiation is emitted. For example, you can feel the heat being radiated by a dish that has just come out of the oven, but the dish does not glow.

> **B** Can you tell if an object is emitting infrared radiation just by looking at it? Explain your answer.

If an object is warmer than its surroundings, it will radiate infrared radiation. If an object is cooler than its surroundings, it will absorb infrared radiation. The hotter something is, the more infrared radiation it will radiate in a given time. If two objects are the same size and one of them is hotter, the hotter one will emit more infrared radiation.

Objects can absorb radiation from the Sun. For example, if you put an object in sunlight, it will absorb infrared radiation from the Sun and its temperature will increase.

The surface of the object affects how much energy it absorbs. A dark matt surface is good at absorbing infrared radiation. A light-coloured or shiny surface is bad at absorbing infrared radiation. This is why the inside of a vacuum flask is shiny.

A dark matt surface is also a good emitter of radiation and a light shiny surface is a poor emitter of radiation.

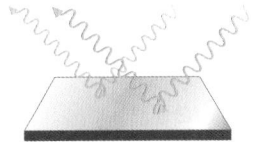
light shiny surface

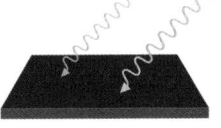
matt black surface

▲ How infrared radiation is absorbed by different surfaces

▲ These buildings are designed to stay cool in the summer as they reflect heat

Energy transfer by infrared radiation

Infrared radiation is a type of electromagnetic wave, like light. It can travel through a vacuum, like light does. It does not need a **medium** to travel through.

Questions

1 Make a table summarising which surfaces are good and which are poor at absorbing and emitting infrared radiation.

2 Why does an electric fire usually have a shiny panel behind the heating element?

3 Look at the picture of the houses. What features help to keep the houses cool in summer?

4 Why is the inside of a vacuum flask shiny?

5 What does a motion detector need to do to set off an alarm?

E

C

A*

Did you know…?

A patio heater is cleverly designed to warm the area around it, whilst minimising energy transfer from the top surface. The top surface is shiny. This reduces the infrared radiation emitted to the surroundings.

▲ This patio heater uses infrared radiation

Exam tip **AQA**

✓ Remember that the hotter an object is, the more infrared radiation it emits.

Learning objectives

After studying this topic, you should be able to:

✔ use kinetic theory to explain the three states of matter

Did you know...?

All big structures have expansion joints. Structures expand and contract as the air temperature changes. For example, if a bridge did not have expansion joints, it could buckle and break in very hot weather.

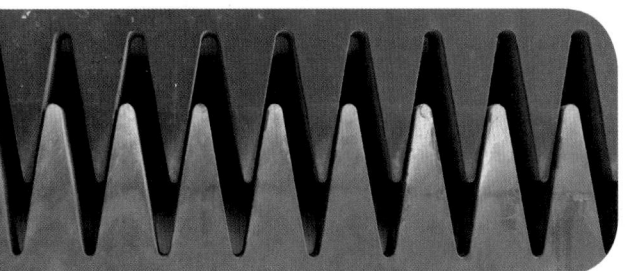

▲ An expansion joint in a bridge

A What are the differences between a solid and a liquid?

B What are the similarities between a liquid and a gas?

States of matter

The three **states of matter** are solid, liquid, and gas. Solids, liquids, and gases are all made up of tiny particles.

In a **solid**, the particles are packed close together in a regular pattern. They are held by attractive forces between them. Each particle vibrates about its fixed position. The shape and the volume of the solid is fixed.

In a **liquid**, the particles still attract one another and are still packed close together, but not in a regular way. They vibrate more and are free to move around. This is why liquids can flow.

In a **gas**, the particles are too spread out to attract one another much. The particles move around freely at high speed and collide with one another and with the walls of the container.

This **kinetic theory** explains the properties of the states of matter. The properties are given in the table.

State	Properties
solid	fixed shape, fixed volume
liquid	no fixed shape, fixed volume
gas	no fixed shape (takes the shape of its container), no fixed volume

States and energy

The particles in solids, liquids, and gases have different amounts of energy.

In gases, the particles have the most energy and are moving around at high speed. Particles in solids are fixed and can only vibrate from side to side. They have less energy than the particles in liquids, which are free to move.

solid

liquid

gas

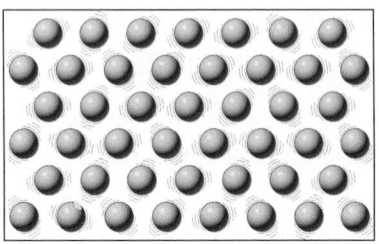
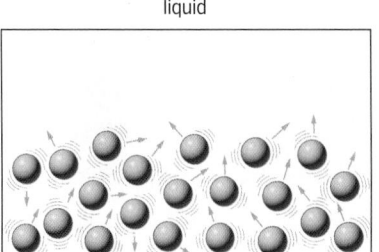
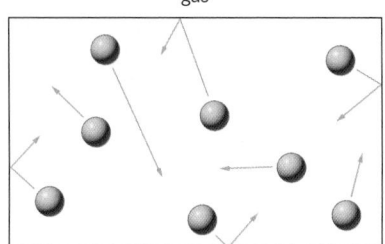

▲ Particles in a solid, a liquid, and a gas. In each case, the amount of energy is related to the temperature of the substance.

As the temperature increases, the size of the vibrations, or the speed of the particles, increases. The particles are taking up more space and the substance expands.

When the temperature decreases, the particles have less energy. In a solid, they vibrate less and so take up less space. The solid contracts.

In liquids and gases, the particles also have less energy and so move more slowly. The gas or liquid becomes more dense as the temperature decreases.

Key words

states of matter, solid, liquid, gas, kinetic theory

C   What happens to the particles in a liquid when the temperature increases?

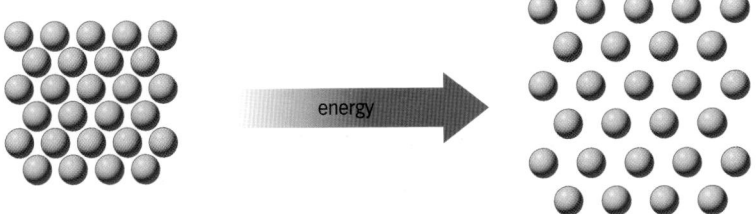

▲ Particles take up more space when the temperature increases

energy

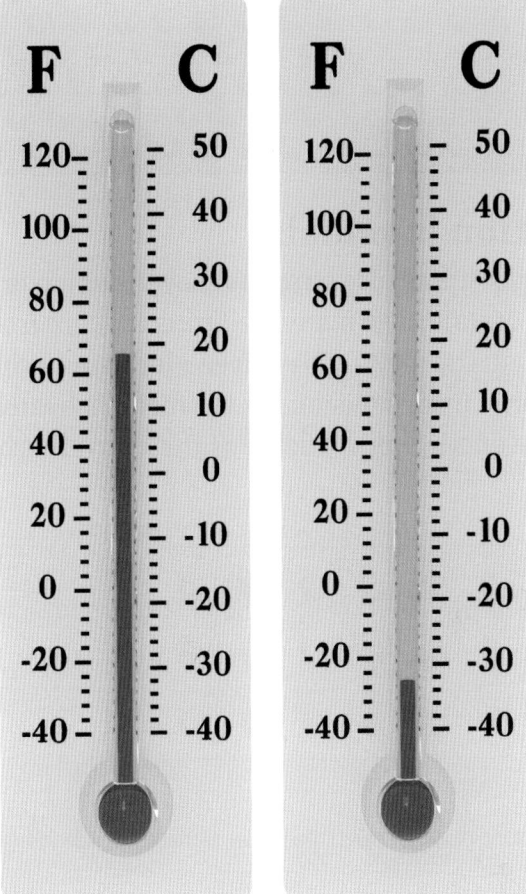

▲ A thermometer uses the principle that a liquid expands when it gets hotter

Questions

1   Make a table summarising how the particles behave in solids, liquids, and gases.   ↓ E

2   Describe what happens to the particles in a gas as the temperature increases.   ↓ C

3   Use the kinetic theory to explain why liquids can flow.   ↓ A*

4   Explain how a thermometer works.

3: Conduction

Learning objectives

After studying this topic, you should be able to:

✔ understand how energy is transferred by conduction

Key words

free electrons, thermal conductor, thermal insulator, conduction

Thermal conduction

Heat energy is transferred through solid materials by conduction. The particles in a solid are always vibrating. When the particles are hotter, they have more energy and so vibrate more.

When you heat one end of an object, the particles start vibrating more. The particles collide with neighbouring particles. Energy is transferred from one particle to another, in the same way that energy is transferred from the cue ball when it collides with another ball on a pool table. Energy can be conducted through the solid by the particles colliding with each other.

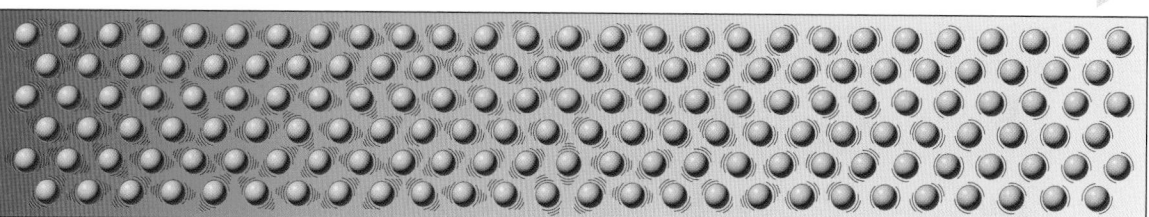

▲ Heat is conducted through a solid

Conduction in metals

In metals something else happens as well, which is more important. Metals have many electrons that are free to move through the metal. These **free electrons** gain more kinetic energy from collisions as the metal is heated. They transfer the energy very quickly as they travel through the metal.

Did you know...?

Diamond is an even better thermal conductor than metals, even though it does not have free electrons. This is because its particles are arranged in a very regular and stable pattern. This also makes it one of the hardest known substances.

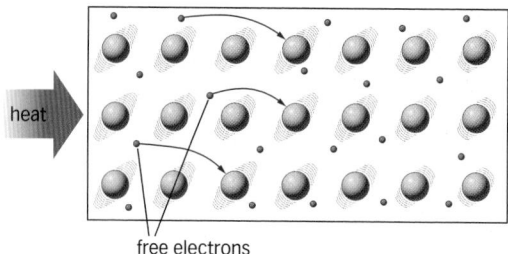

free electrons

▲ Free electrons move through the metal, transferring the energy more quickly

Conductors and insulators

Metals are good **thermal conductors**. Energy is conducted through them easily.

Some materials do not conduct heat very well – they are called **thermal insulators**. Materials such as wood and plastics are insulators and they transfer energy slowly. The particles are close together, but the pattern is not as regular as in a metal. This means that energy only passes slowly from one particle to another.

You can tell whether a material is a conductor or an insulator by touching it. A conductor feels cold when you touch it because energy is conducted away from your skin quickly. Insulators feel warmer because the energy is not conducted away from your skin.

Liquids and gases are poor thermal conductors. The particles in a liquid do not have a regular arrangement, so it is much more difficult for energy to be passed on by **conduction**. In a gas, the particles are far apart so it is a very poor conductor.

A Imagine a metal spoon and a wooden spoon in a metal saucepan over a gas flame. Both spoons are touching the pan. Which spoon will heat up more quickly?

B Describe how the energy gets from the gas flame to the handle of the metal spoon in Question A.

Questions

1 Give an example of a good thermal conductor.

2 An object feels cool when you touch it. Is it likely to be a thermal insulator or a thermal conductor? Explain your answer.

3 Why is the handle of the kettle made of plastic?

▲ The kettle has a plastic handle

4 Why are liquids and gases poor thermal conductors?

5 Explain how free electrons transfer heat in a metal.

Exam tip    AQA

✔ Remember:
a good thermal insulator is a poor heat conductor
a poor thermal insulator is a good heat conductor.

Learning objectives

After studying this topic, you should be able to:

✔ understand how energy is transferred by convection

Key words

fluid, convection, convection current

Did you know...?

Fur works by trapping pockets of air, so that energy cannot be transferred away from the surface of the skin by convection.

This polar bear keeps warm because its fur minimises convection near its skin

Transferring energy in liquids and gases

You have already learned that **fluids** (liquids and gases) are poor thermal conductors. But they can still transfer energy, because the particles are free to move. When moving particles carry energy from one place to another this is called **convection**.

When the particles in one part of a fluid gain more energy, they move faster and that part expands because the particles are taking up more space. There is still the same mass of fluid, but it is taking up a larger volume. The density of that part of the fluid decreases because there is the same number of particles but in a larger space. Its lower density causes it to rise.

The cooler fluid nearby is denser than the heated fluid. The denser cooler fluid falls to the bottom, and the less dense hotter fluid rises. This movement is called a **convection current**.

The diagram shows how a radiator can heat all of the air in a room.

The convection current moves through the room. By the time the air reaches here it will have lost some of its energy and cooled down slightly.

The air particles gain energy from the radiator. The air becomes less dense than the surrounding air and rises.

the cooler more dense air falls

some energy will be absorbed by the walls and objects in the room

▲ Convection currents are used to heat a room

A How is energy transferred from the radiator to the particles in the air?

B What happens to air when it gains energy?

Convection currents

Sometimes you can see the effects of a convection current. For example, when a plant is on a windowsill above a radiator, the convection current set up by the radiator can be strong enough to move the leaves of the plant.

Convection currents can also be deflected. If there is an object in the way, the convection current will flow around it.

Some central heating systems use a convection current to transfer energy from the boiler to the radiators. Convection currents are also used in some domestic hot water systems.

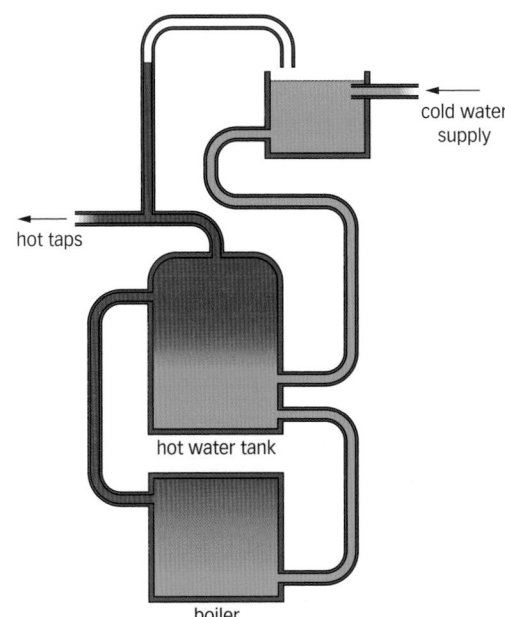

cold water supply

hot taps

hot water tank

boiler

▲ Domestic hot water system

> **C** Why do convection currents occur in liquids?

Questions

1 Draw a diagram and label it to show how convection currents transfer energy through the water in the saucepan shown in the picture.

◀ Saucepan of water being heated on stove

↓ E

2 How is convection used in the hot water system shown in the diagram?

3 Harry is sitting in the chair in a room like the one in the picture on the opposite page, when the heating is switched off on a cold day. Why does he feel cold?

↓ C

4 An ice cube is placed in a glass of water. In terms of energy and particles, explain why the water beneath the ice cube begins to sink.

↓ A*

5: Evaporation and condensation

P1

Learning objectives

After studying this topic, you should be able to:

✔ understand how energy can be transferred by evaporation and condensation

✔ understand how to change rates of evaporation and condensation

Key words

evaporation, condensation

▲ The puddles will dry up because of evaporation

Exam tip AQA

✔ Remember that evaporation has a cooling effect.

Evaporation

The ground doesn't stay wet for very long after rain – it dries by **evaporation**. The kinetic theory helps to explain how this happens.

In a liquid, the particles do not all have the same energy. They have different energies and so some of them will be moving faster than others. Particles with a higher energy than the average are constantly escaping from the surface of the liquid. Some particles fall straight back into the liquid, but others with a higher energy escape into the air. So the total amount of energy in the liquid decreases. As the escaping particles have a higher energy than the average energy of a particle left in the liquid, the average energy of the particles left behind decreases. The temperature of the liquid goes down.

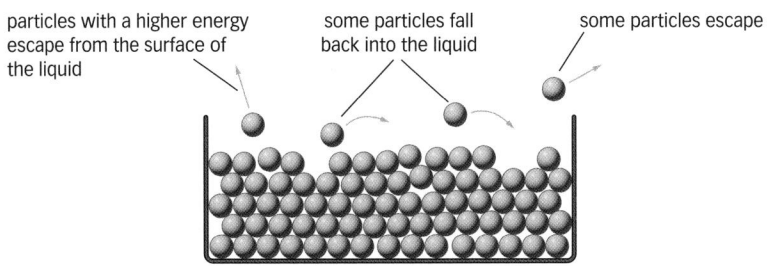

▲ Evaporation from the surface of a liquid

> **A** How does a liquid cool down by evaporation from its surface?

Condensation

Condensation can be explained in a similar way. Particles in a gas have more energy than particles in a liquid. If the particles lose enough energy they will condense to form a liquid.

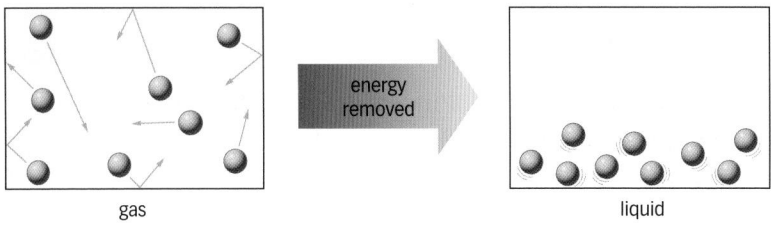

▲ When gas or vapour particles lose energy, they condense into a liquid

Changing the rate of evaporation

You can change the rate of evaporation in these ways:

- Increase the flow of air, for example by blowing a fan over the surface.
- Add energy, by heating the liquid.
- Increase the surface area of the liquid.

How does that work?

Increasing the flow of air means that the air particles that have escaped from the surface of the liquid are carried away before they have a chance to fall back into the liquid.

When you give the particles in the liquid more energy, they have a higher speed and so are more likely to escape from the surface of the liquid.

Increasing the surface area means that more particles are near the surface.

An electric hand dryer dries your hands by blowing air over them. The moving air is also heated, so energy can be transferred to the water on your skin and increase the rate of evaporation.

> **B** How could you change the rate of evaporation?
>
> **C** What are the best conditions for drying washing in the fresh air?

▲ Moisture evaporating from a damp tree in a forest

Questions

1. Why does blowing over a hot drink help to cool it down? ↓ E

2. Explain why you can feel cold when you get out of a swimming pool. Use the kinetic theory in your explanation.

3. Explain how a tumble dryer dries clothes. ↓ C

4. Draw and label a diagram to show how the floating water distiller works.

5. Describe how evaporation can be used to keep things cool, and give an example. ↓ A*

Exam tip

✔ Remember that condensation is the reverse process of evaporation.

Learning objectives

After studying this topic, you should be able to:

- ✔ understand that the shape and size of an object affects how quickly energy is transferred to and from it
- ✔ understand that temperature difference affects how quickly energy is transferred

Exam tip AQA

- ✔ Remember that energy always moves from a warmer place to a cooler place, never the other way round.

▲ What will happen to this cold drink?

Heating and cooling

The hot drink shown in the photo contains energy. The hot drink will cool down because energy is transferred from it to its surroundings. Energy is transferred until the temperature of the drink is the same as its surroundings.

▲ This hot drink will cool down as energy is transferred to its surroundings

Changing the rate

The difference in temperature between an object and its surroundings affects the rate at which energy is transferred. When the temperature difference is greater, the rate of energy transfer is higher.

Size and shape also affect the rate at which energy is transferred. If two objects have the same volume but different shapes, the one with a larger surface area will cool down more quickly.

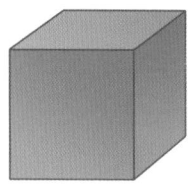 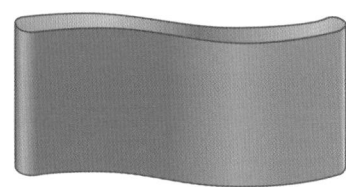

▲ These two objects have the same volume but different shapes

The type of material that something is made of also affects how quickly it transfers heat energy. For example, an object that is made from a good thermal conductor will transfer heat energy more quickly than one that is similar but is made of a thermal insulator.

Transfer of heat energy also depends on what an object is in contact with. For example, you have already learnt that a good conductor feels cool when you touch it because heat energy is quickly conducted away from your hand. Heat energy is transferred much more quickly along a metal spoon than along a wooden one.

metal

wood

▲ Two identical objects – one touching a thermal conductor, the other a thermal insulator

Questions

1 Look at the cold drink shown in the picture. Describe what will happen in terms of the temperature of the drink and the flow of energy. **E**

2 Draw a table to summarise all the things that can change the rate at which heat energy is transferred by heating. **C**

3 How could you reduce the rate of heat energy transfer from a hot drink?

4 What can you say about the rate of heat energy transfer in an object that is at the same temperature as its surroundings? **A***

A Which will cool down to 50 °C more quickly – a kettle of water at 100 °C or a hot drink at 70 °C? Explain your answer.

B Which contains more heat energy – a kettle full of water at 60 °C or a tank full of water at 60 °C?

C Which object in the diagram on the left will cool down more quickly? Explain your answer.

Did you know...?

African elephants live in a hot climate and need to cool down. An elephant has large ears that help to control its body temperature. The ears have a large surface area and so more energy can be transferred to the elephant's surroundings.

▲ African elephants use their ears to control their body temperature

Learning objectives

After studying this topic, you should be able to:

- ✔ compare the ways in which energy is transferred between objects and their surroundings by heating
- ✔ understand how to vary the rate of energy transfer by heating
- ✔ evaluate the design of everyday appliances that transfer energy by heating

Key words

dissipate, evaluate

The cooling fins on a motorbike engine transfer energy to the surrounding air

Sometimes we want to transfer as much energy by heating as possible, for instance with radiators. In other cases we want to minimise energy transfers, for example with double-glazed windows.

Animal adaptations for energy transfer

Some animals in hot countries need to be able to **dissipate** excess energy from their bodies. The cape fox below, for example, has large ears. Blood is pumped through the ears and energy is transferred to them. As the ears have a large surface area, more energy can be transferred to the air. In contrast, the arctic fox has small ears to prevent it losing precious body heat.

▲ A cape fox

▲ An arctic fox

Design for transfer of energy

Central heating radiators are designed to transfer as much energy by heating as possible. Radiators are long and thin so that the surface area is maximised, and some have fins which are bent at the top. This means that the convection current that is caused by the radiator is directed into the room rather than up the wall.

> **A** What feature of the cape fox's ears increases the rate of energy transfer by heating?
>
> **B** What features of the cooling fins in the picture are designed to transfer energy quickly?

A vacuum flask is designed to keep things hot or cold. The transfer of energy by heating is minimised.

The double walls of the glass bottle inside the flask are silvered, which reduces the amount of energy transferred by radiation. There is a vacuum between the two silvered surfaces, so there are no particles to transfer energy by conduction or convection. The glass bottle is held in place inside the outer casing by pads that are made of a poor conductor. The flask also has a tight-fitting screw cap made of a poor conductor (like plastic). This reduces losses by conduction, convection, and radiation.

However, some energy can still be transferred through the vacuum by radiation. Some energy will still be lost through the pads and the cap.

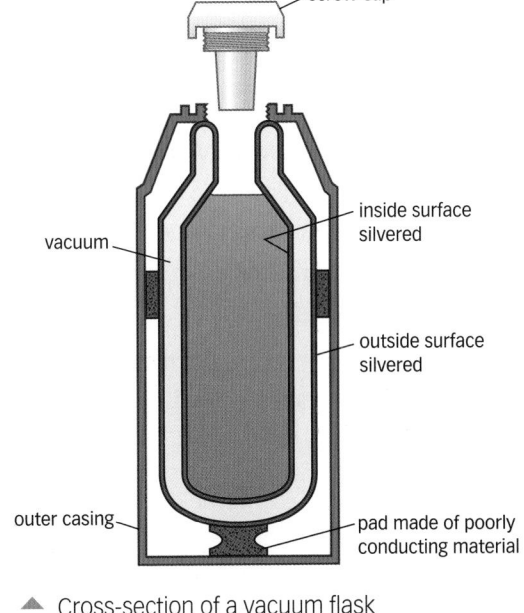

▲ Cross-section of a vacuum flask

Questions

1 How could the engine in the picture be changed to increase the rate of energy transfer?

▲ Video card used in personal computers

2 Look at the picture of the video card. The video card produces energy which needs to be taken away. How do the cooling fins do this?

3 Describe in detail how a vacuum flask keeps a drink cool.

4 What kind of material would you put at the top of the vacuum flask?

5 Why should you fill up a vacuum flask rather than half-fill it?

Did you know...?

Many appliances that dissipate energy will also have a fan to increase the flow of air over them. For example, car engines have a fan that starts automatically in hot weather to increase the flow of air over the radiator when the engine gets too hot.

Exam tip AQA

✔ When you are asked to **evaluate** the energy transfers in an appliance, first look at what the appliance is designed for. Is it to increase or to decrease the rate of energy transfer by heatng? What types of heat transfer might be happening? Then look at the characteristics of the appliance. Does it have a large surface area?

Learning objectives

After studying this topic, you should be able to:

✔ describe how to reduce heat transfer in buildings

✔ understand what a U-value is

✔ understand that better thermal insulators have a lower U-value

Key words

U-value

Did you know...?

Closing curtains when it gets dark in winter can help keep a room much warmer by reducing the energy transferred through the windows.

Insulating houses

Houses transfer energy to their surroundings. It costs money to heat them and so home owners try to reduce the cost of heating by insulating them better. The thermogram below shows that the middle house in the terrace is much better insulated than its neighbours.

▲ The areas transferring the most energy are red (warmest), then yellow, then green, with blue being the coolest

When you insulate a house you reduce energy transfers through conduction, convection, and radiation.

> **A** How do you reduce the energy transfers from a house to its surroundings?

The diagram shows a cross-section of a cavity wall. There are two parts to the wall with a cavity between them. The inner wall is made from a material that does not conduct energy well. If there is nothing in the cavity between the walls, a convection current can be set up which transfers energy from the inner wall to the outer wall. So the cavity is filled with an insulating material. This is usually something with small pockets of trapped air. The trapped air prevents convection currents forming.

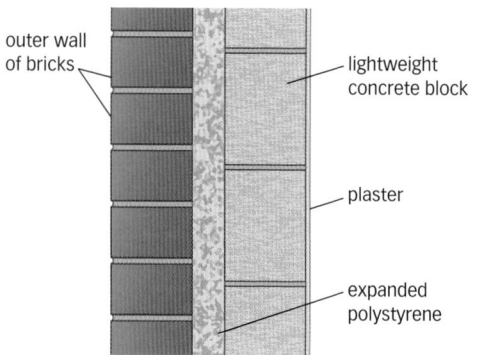

outer wall of bricks

lightweight concrete block

plaster

expanded polystyrene

▲ Cross-section of a cavity wall

U-values

Different parts of a building will conduct energy out of a building at different rates. For example, the windows in your house will usually transfer energy more quickly than the walls. We can measure the rate of losing energy for each part of a building. This is called its **U-value**. The higher the U-value, the more quickly energy is transferred. A part or a material with a low U-value is a good insulator.

House builders try to use materials with the lowest U-values. You can also reduce U-values by adding more insulation in the roof or by using glass that transfers less energy.

Part of building	U-value
270 mm cavity wall, no insulation	1.0
270 mm cavity wall with insulation	0.6
single-glazed window	5.0
double-glazed window	2.9
roof material, 50 mm insulation	0.6
roof material, 100 mm insulation	0.3

▲ Some typical U-values found in a building

Questions

1 Describe how the double-glazed window reduces the transfer of energy.

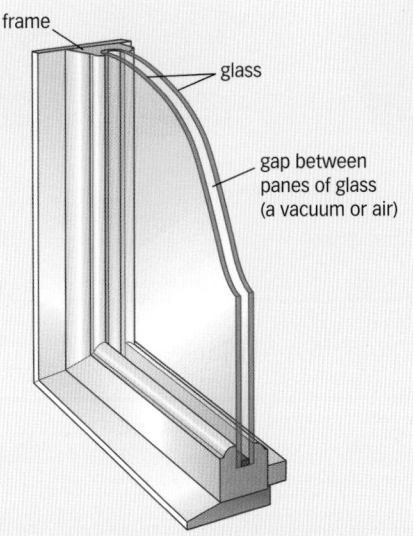

▲ Cross section of a double-glazed window

2 Look at the table of U-values, Which walls, roof and windows would you use to reduce the amount of energy needed to keep the house warm?

3 People often put a shiny covering on the wall behind a radiator. Explain why they do this.

4 Look at the thermogram of the house on the previous page. What could be causing the red spot on the roof of the middle house?

B Look at the table above. Which part of the building is insulated best?

Exam tip

✓ You might be asked to evaluate features of a house by looking at the U-values. Remember that a better insulator has a lower U-value.

Solar panels

Solar panels absorb infrared radiation from the Sun. The energy is transferred from the panel to the metal pipes in the panel, and from there to the water in the pipes.

Usually, a pump is not needed to move the water through the system. A natural convection current occurs in the system which transfers the energy to the hot water tank.

When the Sun is shining on a cold day, there is still infrared radiation from the Sun. This means that solar panels can still heat water when the outside temperature is below freezing on a sunny day.

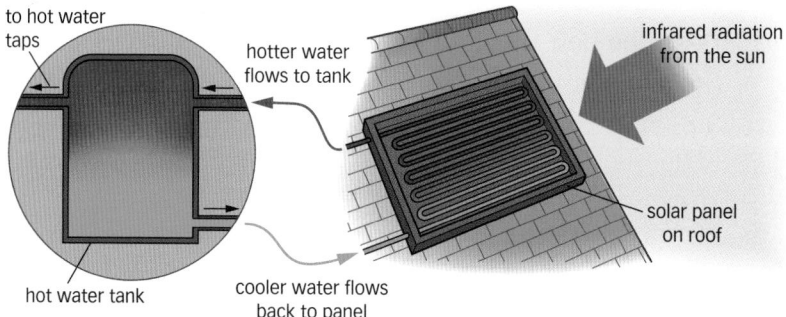

to hot water taps

hotter water flows to tank

infrared radiation from the sun

solar panel on roof

hot water tank

cooler water flows back to panel

▲ How a solar panel works

A Why is a solar panel black?

▲ Solar panels on a roof

Did you know...?

Solar panels can still heat water on a cloudy day. But they don't transfer as much energy to the water as they do on a sunny day.

Payback time

You can cut down your energy bills by reducing the amount of energy that you use. Some ways of doing this do not cost anything. For example, you could turn the thermostat on your central heating down by 1 °C or draw the curtains when it gets dark.

Other methods do cost money. To decide whether they are worth doing, you need to calculate how long it will take to recover the amount of money you spend. This is the **payback time**. The method with the shortest payback time is the most **cost-effective**.

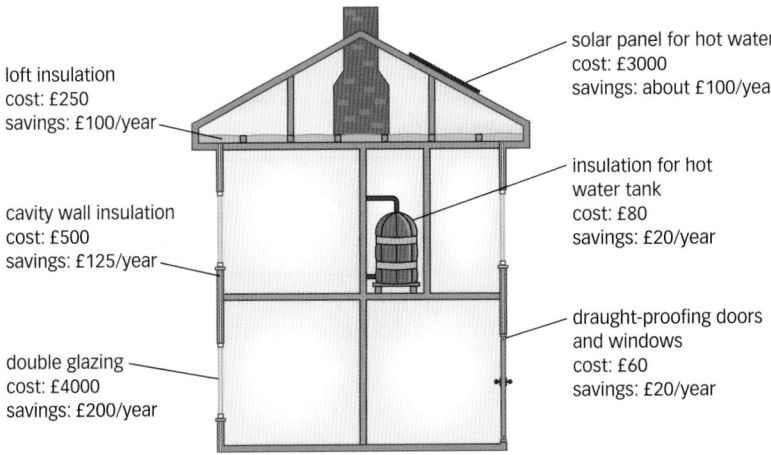

loft insulation
cost: £250
savings: £100/year

cavity wall insulation
cost: £500
savings: £125/year

double glazing
cost: £4000
savings: £200/year

solar panel for hot water
cost: £3000
savings: about £100/year

insulation for hot water tank
cost: £80
savings: £20/year

draught-proofing doors and windows
cost: £60
savings: £20/year

▲ Different ways of reducing your energy consumption

Worked example

Jack currently has 50 mm insulation in his loft. It will cost £250 to increase the thickness of insulation to the recommended 270 mm. His heating bill would decrease by £100 per year. What is the payback time for adding the extra insulation?

$$\text{payback time} = \frac{\text{cost}}{\text{savings per year}} = \frac{£250}{£100 \text{ per year}} = 2.5 \text{ years}$$

Exam tip

✔ You do not need to remember any particular values for payback calculations. In exam questions, you will be given the data and asked to calculate the payback time to evaluate which methods are the most cost-effective.

B What is payback time?

Questions

1 Describe all the energy transfers that take place in a solar panel system. ↓E

2 Look at the diagram of the house. Which method of reducing energy consumption is:

(a) most cost-effective?

(b) least cost-effective?

3 Sometimes you can get grants to help with the cost of some kinds of insulation. Work out the payback time for: ↓C

(a) solar panels, when there is a grant of £500

(b) loft insulation, when there is a grant of £200.

4 What is the ideal direction for a solar panel to be facing in the UK? ↓A*

Learning objectives

After studying this topic, you should be able to:

✔ understand the idea of specific heat capacity of a material

✔ use specific heat capacity to work out how much energy is needed

A What is meant by the specific heat capacity of a material?

▲ How much energy is required to heat the oil to fry food?

B What energy transfer is needed to heat a 500 g aluminium pan containing 1 kg of cooking oil from 20 °C to 170 °C?

C The amount of energy you have calculated is less than what would actually be needed. Explain why.

The amount of energy an object stores is related to its mass and temperature. It also relates to the material the object is made of. For example, it takes more energy to raise the temperature of 1 kg of water than it does to raise the temperature of 1 kg of aluminium.

The **specific heat capacity** of a material is the amount of energy needed to raise the temperature of 1 kg of the material by 1 °C. Its units are J/kg °C.

Some specific heat capacities are given in the table below.

Material	Specific heat capacity (J/kg°C)
water	4200
aluminium	880
copper	380
cooking oil	1200 (about)

You can use this equation to calculate the amount of energy needed to heat an object:

$$E = m \times c \times \theta$$

E energy transferred (joules, J)	=	m mass (kilograms, kg)	×	c specific heat capacity (J/kg°C)	×	θ temperature change (degrees Celsius, °C)

Calculating the amount of heat transferred

Worked example

A kettle contains 1.5 kg of water at a temperature of 18 °C. How much energy is needed to bring the water to the boil?

energy needed = mass × specific heat capacity of water × temperature change

= 1.5 kg × 4200 J/kg °C × (100 − 18) °C

= 1.5 × 4200 × 82 J

= 516 600 J = 516.6 kJ

Calculating specific heat capacity

You can work out the specific heat capacity of a material if you know the amount of energy transferred, the mass of the object and the change in temperature.

$$\text{specific heat capacity (J/kg}^\circ\text{C)} = \frac{\text{mass (kg)}}{\text{temperature difference (}^\circ\text{C)}}$$

Key words

specific heat capacity

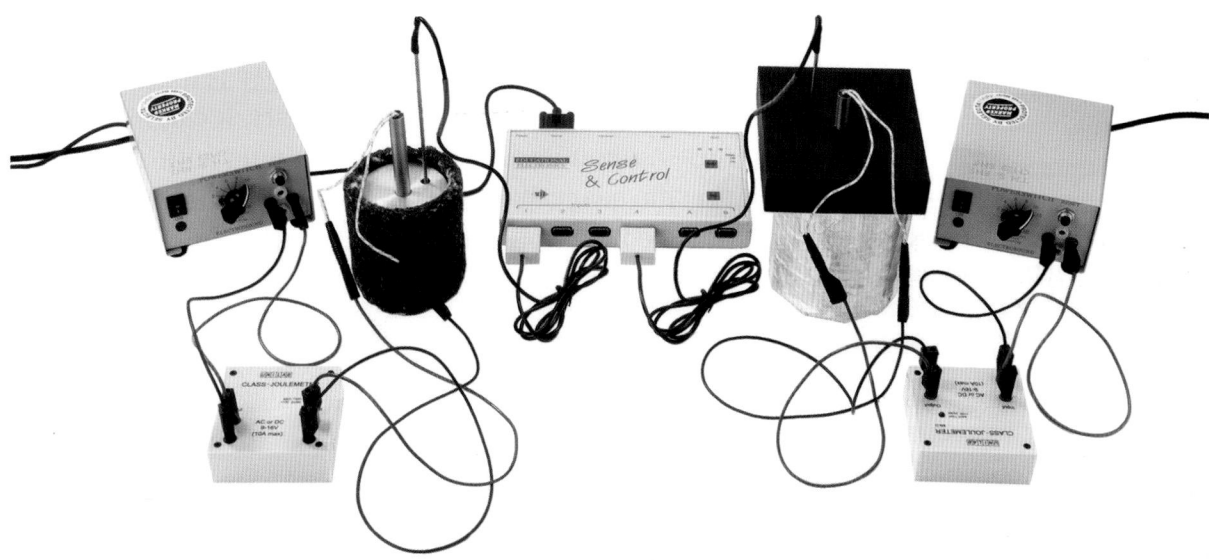

▲ This apparatus is being used to find the specific heat capacity of the material in a metal block

Questions

1 What are the units of specific heat capacity?

2 Look at the apparatus in the picture above. Why is the metal block wrapped in an insulating cover? Explain your answer.

3 Which needs more energy: to heat 5 kg of copper from to 20 °C to 50 °C or to heat 0.5 kg of water from 20 °C to 50 °C?

4 50 kJ of energy was transferred to a material with a mass of 5 kg. The temperature increased from 20 °C to 60 °C.
What is the specific heat capacity of the material?

▼ E

↓ C

↓ A*

Exam tip AQA

✔ Don't forget to work out the temperature difference when you are calculating how much energy is needed.

Learning objectives

After studying this topic, you should be able to:

✔ understand how materials with different specific heat capacities are used

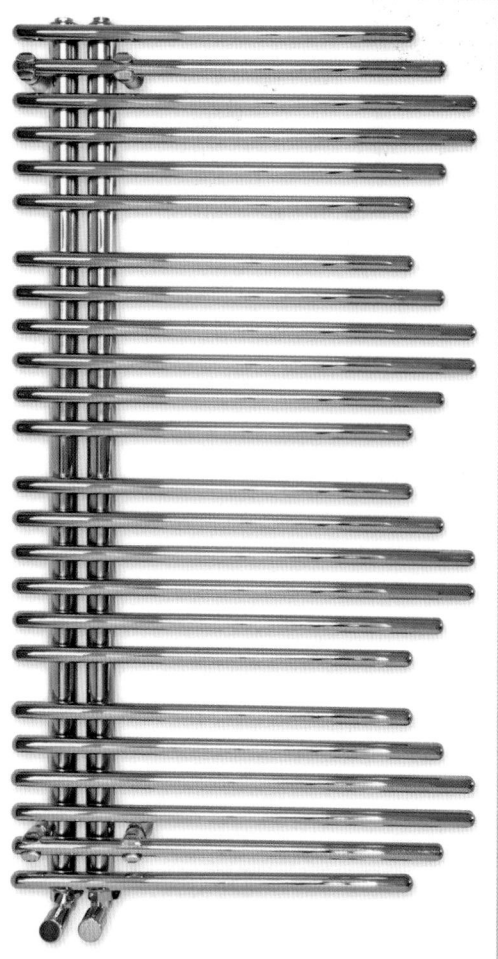

▲ Water is used to transfer energy from the boiler to radiators like this one

Using different materials

You know that the specific heat capacity of a material tells us how much energy it can store. Different materials are used for particular purposes because of their specific heat capacities.

Water is used in central heating systems to transfer energy from the boiler to the radiators. Water is used because it has a very high specific heat capacity. It can transfer much larger amounts of energy than a liquid with a lower specific heat capacity. This also means that the water does not need to be pumped very quickly around the central heating circuit.

> **A** Why is water used in central heating systems?

Electric storage heaters contain blocks of concrete or bricks that store energy. The heater uses cheap electricity during the night to heat the blocks. The energy is released slowly during the day.

Concrete has a lower specific heat capacity than water. It is used because it is simpler to make the heaters than if water is used and is easier to maintain. The storage heater can store a lot of energy because it contains large blocks of concrete. This also means that the temperature of the blocks can be lower than if less material was used.

▲ This heater contains blocks of concrete or bricks that store energy

Oil-filled radiators can also store energy. They are designed to provide a gentle steady heat. They do this by using oil which has a high specific heat capacity. Lots of energy is stored in the oil and then radiated and carried by convection from the surface of the heater.

Material	Specific heat capacity (J/kg°C)
water	4200
concrete	880
oil	1500
glass	500–840
wood	1700

▲ Specific heat capacities of some materials

▲ This heater is filled with oil which has a high specific heat capacity

B Why does a storage heater need to store lots of energy?

Questions

1 Water is used to cool car engines. Explain why water is used and not some other substance.

2 In hot countries, some buildings have very thick walls. How does this help to keep the inside of the building cool?

3 Wood has a high specific heat capacity. Why would you not use it in a heater?

4 Alex suggests replacing the oil in the heater with a material that has a much lower specific heat capacity. What changes would you have to make to the heater?

↓ E

↓ C

↓ A*

Did you know...?

Some office buildings use blocks of concrete to help keep them cool in hot weather. The ceiling of each storey is a large block of concrete. It absorbs excess heat during the day. It then cools down by transferring the energy back into the air at night, when there is no-one in the building.

Exam tip

✔ You need to be able to apply the principles of specific heat capacity. In the exam, you may be given one of these examples or a completely new example.

Learning objectives

After studying this topic, you should be able to:

✔ describe energy transfers

✔ understand that energy is not created or destroyed, only transferred

Key words

kinetic energy, chemical energy, nuclear energy, elastic potential energy, gravitational potential energy, transfer, useful energy, wasted energy, law of conservation of energy

As well as heat, energy takes many forms such as light, sound, **kinetic energy**, and electrical energy.

Energy can be stored in different ways:

- **chemical energy** is stored inside food, fuels and batteries
- **nuclear energy** is stored inside the nucleus of atoms
- **elastic potential energy** is stored in anything that is stretched or squashed like a stretched rubber band or a coiled spring
- **gravitational potential energy** is stored in any object that is higher than its surroundings.

▲ All of these store chemical energy

You have already seen how energy can be transferred by conduction, convection, and radiation. Many appliances or machines **transfer** energy from one form to another. For example:

- a kettle transfers electrical energy into heat energy and sound energy
- a petrol mower transfers chemical energy stored in petrol or diesel into kinetic energy, heat energy, and sound energy.

▲ This wind-up toy stores elastic potential energy

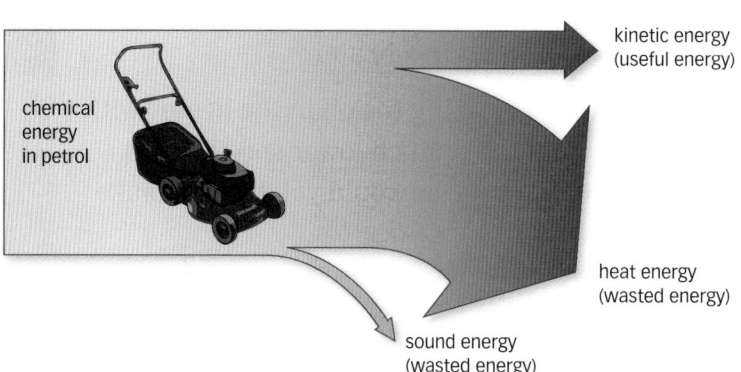

chemical energy in petrol

kinetic energy (useful energy)

heat energy (wasted energy)

sound energy (wasted energy)

▲ Energy transfers in a petrol mower

A What are the energy transfers when you release a stretched spring?

B What energy transfers happen when you light a barbecue?

Useful and wasted energy

Machines usually transfer energy into more than one form. Often only one of these transfers is useful. For example, when a kettle transfers electrical energy into heat energy, this is a **useful energy** transfer. The sound energy is not useful – this is **wasted energy.**

C In Questions A and B, which energy transfers are useful and which ones produce wasted energy?

You can reduce your energy consumption by using appliances that transfer less energy into wasted forms. For example, you could replace old light bulbs with low-energy bulbs which transfer much less energy into wasted heat energy.

Modern cars also transfer less energy than older ones into wasted heat energy.

Conservation of energy

The total amount of energy before and after these energy transfers is always the same. All the different forms of energy output from the mower shown in the diagram on the left add up to the amount of chemical energy that was supplied to the mower.

Energy cannot be created or destroyed. It can only be transferred from one form to another. This is the **law of conservation of energy**.

◀ LED light bulbs

Questions

1 What energy transfers take place in the wind-up toy pictured?
↓ E

2 What energy transfers take place in:
   (a) gas-powered hair curling tongs?
   (b) a wind-up radio?
   In each case state which of the transfers are wanted and which are unwanted.
↓ C

3 A computer transfers 120 J of energy every second.
   (a) Write down the energy transfers that are taking place.
   (b) What is the total energy output per second in all these forms of energy?

4 An energy-saving light bulb uses 11 W and produces the same amount of light as a 75 W filament light bulb. Why should you use the energy-saving bulb instead of the filament bulb?
↓ A*

Did you know...?

You can now buy LED lights that produce the same amount of light energy as energy-saving light bulbs, but use even less electrical energy.

Learning objectives

After studying this topic, you should be able to:

- ✔ understand more about useful and wasted energy
- ✔ understand that energy becomes increasingly spread out
- ✔ understand what the efficiency of an appliance is
- ✔ calculate the efficiency of an appliance

Key words

efficiency

As a hot drink cools, energy is transferred to the air around it. The surroundings become warmer but you may not notice this. The energy spreads out further and further. As this happens, it becomes more difficult to use this energy for further transfers.

▲ This hot drink transfers energy to its surroundings

Efficiency

You have already seen that when a machine transfers energy, only some of the transferred energy is useful and the rest is wasted.

For example, a light bulb transfers electrical energy into heat as well as light. Light is useful energy, but the heat is wasted energy. The Sankey diagram shows these transfers in a light bulb.

▲ An energy-efficient light bulb (centre left), two conventional filament bulbs, and a halogen bulb (centre right)

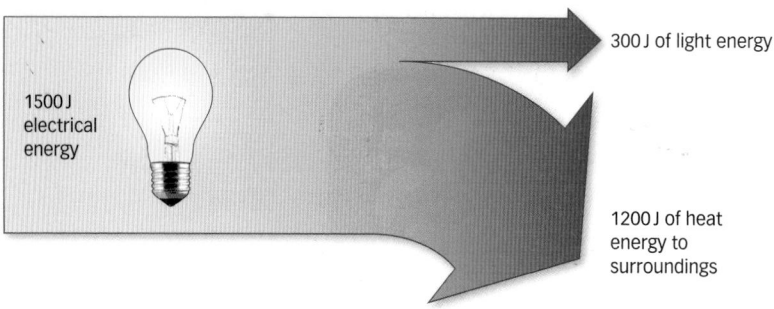

1500 J electrical energy

300 J of light energy

1200 J of heat energy to surroundings

▲ Sankey diagram for a light bulb

> **A** What is meant by the efficiency of an appliance?

We can work out how efficient the light bulb is, using

$$\text{efficiency} = \frac{\text{useful energy transferred}}{\text{total energy supplied}}$$

Remember that the units for energy transferred and energy supplied should be the same – they should both be joules or both in kilojoules.

You can also give the efficiency as a percentage by multiplying the answer by 100%.

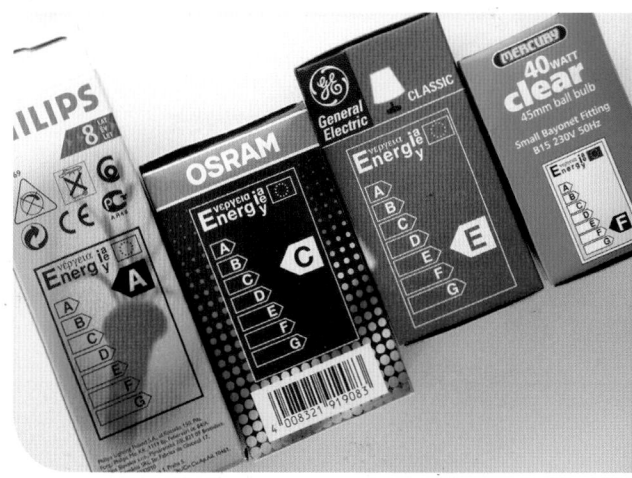

▲ Some of these light bulbs transfer electrical energy into light energy much more efficiently than the others

Worked example

What is the efficiency of the light bulb shown in the diagram?

useful energy transferred = 300 J

total energy supplied = 1500 J

$$\text{efficiency} = \frac{300}{1500} = 0.2 \text{ or } 20\%$$

Did you know...?

All electrical appliances must now have an energy efficiency label. The label shows how much energy you would expect to use in a year. It also grades the appliance in one of seven categories, from A to G.

Questions

1 Which label in the picture on the right shows the most efficient light bulb?

2 A kettle is supplied with 500 kJ of electrical energy. It transfers 400 kJ of heat energy to the water in it, 99 kJ of heat energy to the kettle itself and its surroundings, and 1 kJ into sound energy.

 (a) What is the efficiency of the kettle?

 (b) How could you make the kettle more efficient?

 (c) Draw a Sankey diagram to show the energy transfers that are happening in the kettle.

3 A halogen light bulb transfers 250 J of electrical energy into 220 J of heat energy and the rest into light energy. What is its efficiency?

Exam tip AQA

✓ Remember that efficiency can never be greater than 100%. If your calculations produce an efficiency of greater than 100%, go back and check them – you will have done something wrong!

✓ If you write efficiency as a ratio, it can never be greater than 1.

Learning objectives

After studying this topic, you should be able to:

✔ give examples of energy transfers in electrical appliances

✔ understand that the amount of energy used depends on the appliance's power and how long it is switched on

Key words

joule, power, watt, kilowatt, power rating

Energy transformations

We use electrical energy a great deal in our everyday lives, because it can readily be transferred into other types of energy.

Each electrical appliance is designed to bring about energy transfers. Some of these will be unwanted. For example, a toaster produces heat energy, but it will also produce some light energy (from the glowing elements).

Some appliances may be designed to produce more than one energy transfer. For example, an MP3 player transfers electrical energy into sound energy. It also transfers electrical energy into light energy, so that you can see which track is being played.

▲ These appliances all transfer electrical energy into other forms of energy

Different types of electrical appliances have advantages and disadvantages. For example, a battery-powered radio will run for a long time, but eventually you need to replace the batteries. A clockwork radio does not need batteries, but you need to wind it up regularly.

▲ A traditional (top) and a clockwork (bottom) radio. A clockwork radio offers a number of advantages over a traditional battery-powered one, but is not without its own problems.

Power

You have already learned that energy can be measured in **joules** (J). The amount of energy an appliance transfers per second is called its **power**. An electric blanket and an electric heater both transfer electrical energy to heat energy, but the electric heater transfers many more joules of energy per second than the electric blanket. The power of the heater is higher than the power of the blanket.

Power can be measured in joules per second or in **watts** (symbol W). 1 joule per second is the same as 1 watt. 1000 W is 1 **kilowatt** (symbol kW).

Many electrical appliances have labels showing how much power is needed to run them. This is called the **power rating** of the appliance.

The total amount of energy transferred by an appliance depends on how long it is switched on for, as well as its power. You can work out how much electrical energy an appliance transfers:

$$\begin{array}{ccccc} \text{energy} \\ \text{transferred} & = & \text{power} & \times & \begin{array}{c}\text{how long the appliance} \\ \text{is switched on for}\end{array} \\ E & = & P & \times & t \end{array}$$

▲ Power rating label on a heater

Exam tip   AQA

✔ In an exam, when you are asked to compare different electrical appliances, the question will give you all the data you need.

Worked example

A kettle takes 3 minutes to boil some water. The power of the kettle is 3 kW. How much energy is transferred by the kettle?

3 minutes = 3 × 60 = 180 seconds

3 kilowatts (kW) = 3000 watts (W)

energy transferred = power × time

= 3000 W × 180 seconds

= 540 000 J (or 540 kJ)

A The rating of the heater in the photo is 2000 W. What is this in kilowatts?

B The power rating of a computer is 125 W, and it is switched on for eight hours. The power rating of a toaster is 1.2 kW, and it is switched on for ten minutes. Which appliance transfers more energy?

Questions

1 What is the power of an electrical appliance?

2 Draw up a table and list the appliances shown in the photo on the previous page. List all the useful energy transfers and wasted energy transfers that each appliance makes.

3 What are the advantages and disadvantages of mains-operated fans and battery-operated fans?

4 List all of the things you could not do in your everyday life:

(a) if you could only use appliances with batteries

(b) if you could not use any electrical appliances.

Learning objectives

After studying this topic, you should be able to:

✔ work out the amount of energy transferred from the mains supply by an appliance

✔ work out the cost of energy transferred from the mains supply

Key words

kilowatt-hour, unit

Exam tip **AQA**

✔ If power is given in watts, don't forget to convert it to kilowatts by dividing by 1000 when you substitute it into the equation.

A A light bulb uses 20 W and is switched on for ten hours. How much energy does it use? Give your answer in kWh

Energy transferred

You have to pay for the amount of electricity that you use, but how can you work this out? You have already learned that the total amount of energy used depends on the power of the appliance and how long it is switched on for. Or we can say:

$$E = P \times t$$
$$\text{energy transferred} = \text{power} \times \text{time}$$

If the power is given in kilowatts (kW), and the time is in hours, then the amount of energy used is measured in **kilowatt-hours (kWh)**:

$$\text{energy transferred (kWh)} = \text{power (kW)} \times \text{time (hours)}$$

On your electricity bill, kilowatt-hours are called **units** of electricity.

You can also calculate the amount of energy used in joules. The power is given in watts, and the time is given in seconds:

$$\text{energy transferred (J)} = \text{power (W)} \times \text{time (s)}$$

Worked example 1

A computer uses 250 watts and is switched on for 5 hours. How much energy does it use?

$$\text{energy transferred} = \text{power} \times \text{time}$$
$$= (250 \div 1000) \text{ kW} \times 5 \text{ hours}$$
$$= 1.25 \text{ kWh}$$

▲ Light bulb switched on

Cost of energy transferred

The amount of electricity you use at home is recorded by an electricity meter. Electricity is charged by the unit, or kilowatt-hour. You can work out the cost of the electricity that an appliance uses if you know how much energy it transfers.

The cost of electricity used is given by the equation:

cost = energy transferred × cost per unit

▲ An electricity meter

Worked example 2

A kettle transfers 3.5 kWh of energy. A unit of electricity costs 10.2p. What is the cost of the energy transferred by the kettle?

cost of electricity = energy transferred × cost per unit

= 3.5 kWh × 10.2p/kWh

= 35.7p

B What is the cost of the energy transferred by the light bulb in Question A? Assume that a unit of electricity costs 10.2p.

▲ This meter shows how much the electricity you are using costs

Questions

1 What is a kilowatt-hour?

2 An electric heater has two settings. The first setting uses 800 W and the second 1.5 kW. Work out the cost of using the fire:

(a) for five hours at the 800 W setting

(b) for two hours at the 1.5 kW setting.

Assume that a unit of electricity costs 10.2p.

3 The meter in the photo shows the reading on 31 July. The reading on 31 October is 09231. The first 250 kWh are charged at 16.5p and the remainder at 10.2p.

Work out the cost of the electricity bill for this period.

4 A kettle has a power of 3 kW. It takes three minutes to boil when it is full. A unit of electricity costs 10.5p. What is the cost of boiling a full kettle?

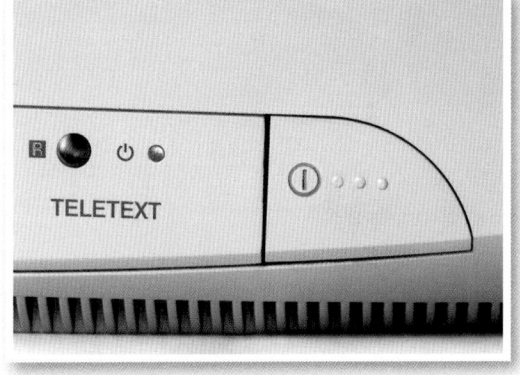

▲ This TV is in standby mode

Did you know...?

When an electrical device is in standby mode, it still uses power. The only way to stop it using any power is to switch it off at the wall socket.

Course catch-up

Revision checklist ✔

- The transfer of energy from a hot object to a cool object by electromagnetic waves is infrared radiation. Dark matt surfaces emit and absorb infrared radiation better than light shiny surfaces.

- The three states of matter, solids, liquids, and gases, are explained by the kinetic theory of particles.

- Metals conduct energy well due to free electrons. Some solids do not conduct energy very well and are thermal insulators.

- Liquids conduct energy, their particles are free to move. This is convection. Convection currents develop when denser cool fluid falls and less dense warmer fluid rises.

- Particles in liquids have different amounts of energy. Energetic particles escape from the surface. When these particles lose energy condensation occurs. Evaporation is increased by giving the liquid more energy.

- Energy transfers from a hot object to its surroundings at a greater rate if the temperature difference is greater. A larger surface area and volume increase the rate of transfer.

- Rate of energy transfer also depends on the material from which an object is made, and the nature of the surface with which it is in contact. A vacuum flask is designed to minimise energy transfer.

- Animals in hot countries are adapted to dissipate energy quickly, unlike those in cold countries.

- Buildings lose energy by conduction, convection and radiation. Insulation helps to prevent this. The rate of energy loss of a material is called its U-value.

- Solar panels provide warm water when infrared radiation from the sun is transferred to water in pipes. Solar panels and insulation keep heating costs down.

- The amount of energy required to change the temperature of one kilogram of the substance by one degree Celsius is its specific heat capacity (SHC). It is different for all substances.

- Water has a high SHC and is useful in central heating systems. Concrete blocks are also used, but have a lower SHC.

- Energy has many forms: kinetic, chemical, light, sound, elastic and gravitational potential, nuclear, and electrical. Energy is neither created nor destroyed.

- Calculating the efficiency of energy transfers within appliances can help to prevent energy being wasted as heat and sound.

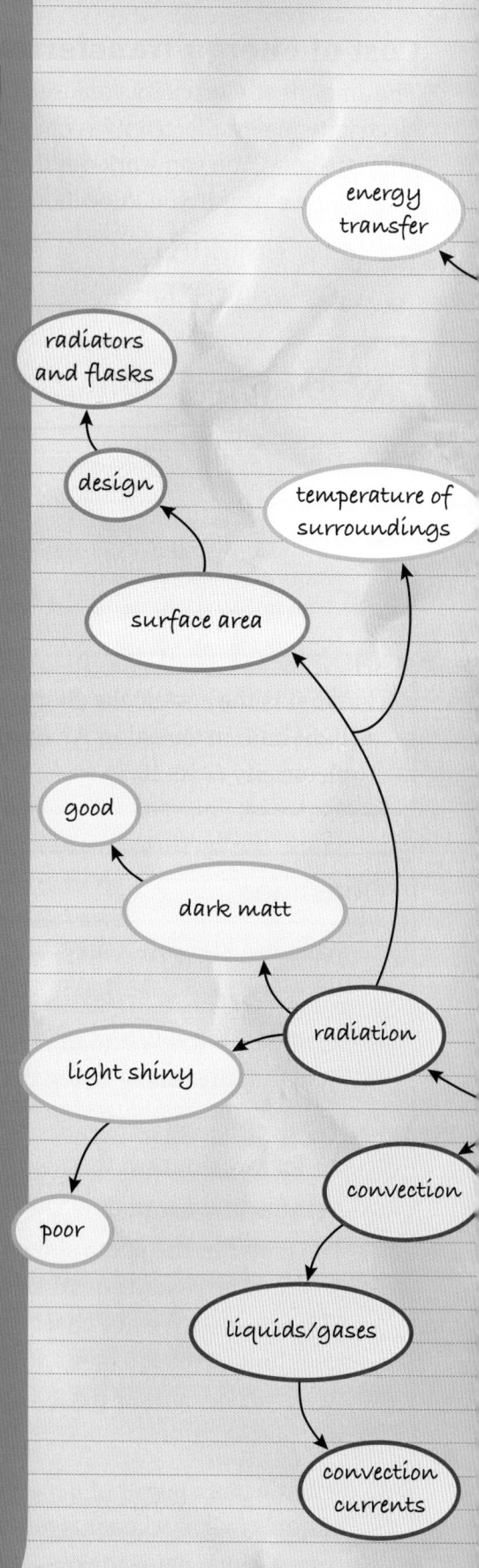

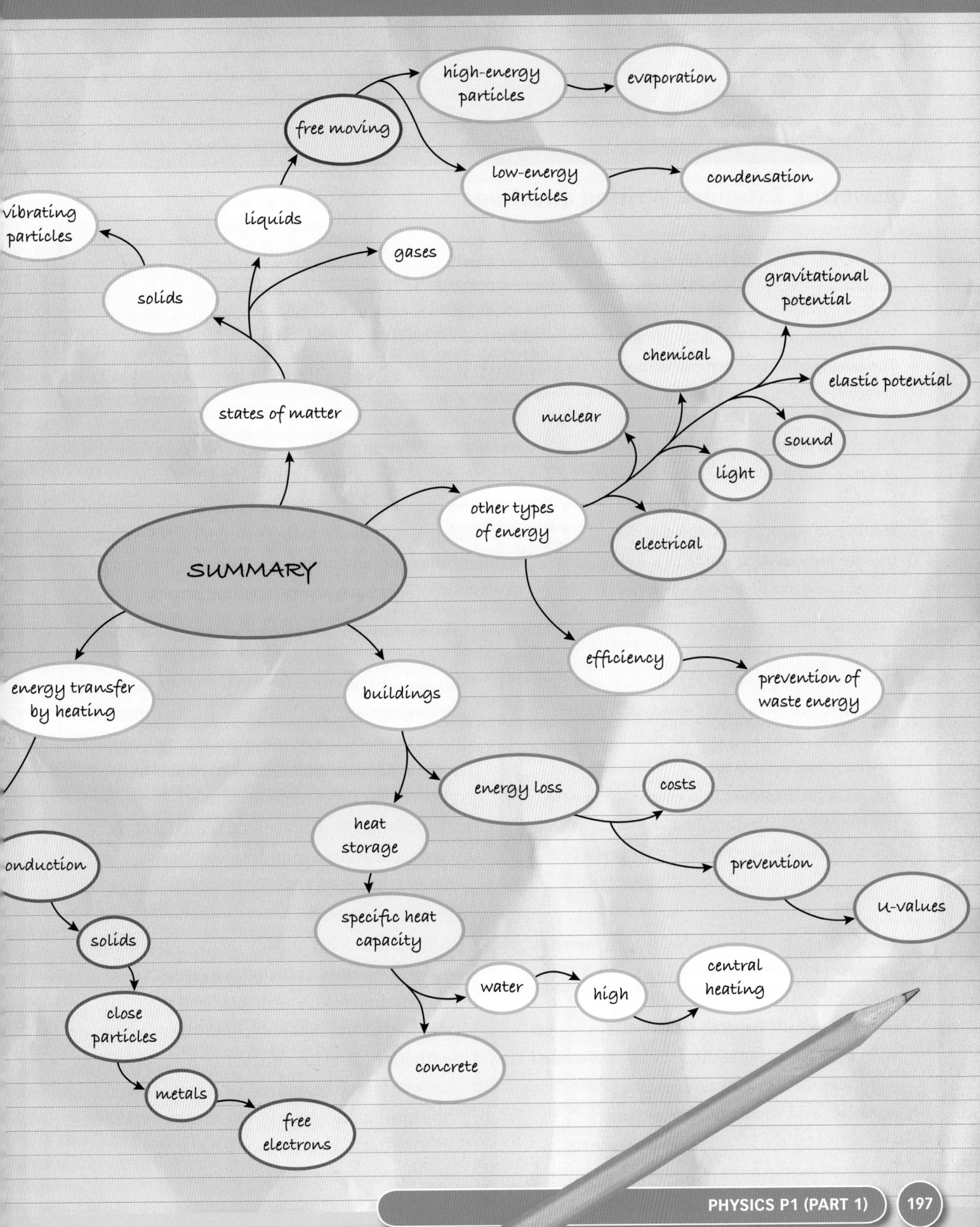

high-energy particles

evaporation

free moving

low-energy particles

condensation

liquids

vibrating particles

gases

solids

states of matter

gravitational potential

chemical

nuclear

elastic potential

sound

light

other types of energy

electrical

SUMMARY

efficiency

prevention of waste energy

energy transfer by heating

buildings

energy loss

costs

heat storage

prevention

conduction

specific heat capacity

u-values

solids

water

high

central heating

close particles

concrete

metals

free electrons

AQA Upgrade

Answering Extended Writing questions

QUESTION

Dave wants to insulate his loft. He looks up the U-values for 100 mm thickness of three possible insulation materials: sheep wool – 0.39 W/m²K; recycled plastic – 0.42 W/m²K; fibreglass – 0.43 W/m²K.

Outline how Dave can use the U-values to help him choose between the three insulation materials. As well as the U-values, what other factors might Dave consider in making his choice?

The quality of written communication will be assessed in your answer to this question.

G–E

Fibreglass is best because it has a hi U-value. It is cruel to use sheep wool for insulation. Recycled plastic is best becuase no much carbon diokside goes into the air when it is made, so making it does not make globel warming worse.

Examiner: The candidate makes a good point about the environmental benefits of using recycled plastic. However, the candidate is wrong to think that materials with higher U-values are better insulators. No credit is given for the comment about wool because it needs more explanation. There are several spelling errors.

D–C

The U-values are nearly the same, but sheep wool have a slightly lower value, so is a little better. Producing wool causes global warming, because sheep breathe out carbon dioxide, the plastic is good because it is recycled. You need to wear gloves with fibreglass because it has sharp bits in it, the others are safer to put in.

Examiner: The candidate understands that materials with lower U-values are better insulators, but has not used the terms 'insulator' or 'insulation'. The point about producing sheep wool is well explained, but the comment about plastic needs more explanation. The spelling is good, but there are errors in grammar and punctuation.

B–A*

The U-value shows how quickly each material transfers heat. Based on U-values, Dave should choose the sheep wool, because it's U-value is lowest. He should also consider payback time. This is the time it takes to get back the money he spends. So he needs to know the price of each material, and what savings to expect per year. He could also consider the environmental impact of making the insulation.

Examiner: This answer includes the main scientific points. It includes a clear explanation of the term 'payback time', and shows that the candidate knows that materials with lower U-values are better insulators. The answer is well organised, and includes only one grammatical error. The answer would be even better if the candidate had given more detail about environmental impacts.

Exam-style questions

B-A*

1 Match each description with the correct energy type below, and write each pair out in a list.

A01

Energy type	Description
heat	stored inside fuel, food, and batteries
light	stored inside atoms
sound	energy stored inside a stretched rubber band
electrical	stored energy because of its height above the ground
nuclear	produced by visible electromagnetic radiation
chemical	energy due to movement of an object
kinetic	produced from a drum, guitar, speaker, etc
gravitational potential	energy transferred by electrons in a conductive wire
elastic potential	energy produced when particles vibrate rapidly

G-E

2 Match up the sentences showing which method of energy transfer is prevented by which means in a vacuum flask.

A03

Method of energy transfer	Means of preventing energy transfer
conduction	silvered glass walls
convection	tight fitting plastic screw cap
radiation	vacuum within the glass layers
	inner casing separated by air gap to outer casing

D-C

3 Draw Sankey diagrams on graph paper using the data for the following machines which are doing work:

A01

B-A*

Machine	% efficiency	% wasted energy
car petrol engine	15	60 – heat 10 – sound 15 – moving parts
fossil fuel electrical power station	35	45 – heat 10 – moving parts 10 - sound
train diesel engine	36	45 – heat 10 – moving parts 9 – sound

Extended Writing

4 Isabelle and Reegan both noticed that after a hot summer's day the bricks of their houses gave off a lot of energy during the evening.

A02

Why does this happen?

G-E

5 Tom is trapped on a small desert island with no fresh water to drink. He remembers from his science lessons that he may be able to evaporate sea water to get pure drinking water. What could Tom do to ensure that he has a fresh supply of water?

A03

D-C

6 The diagram below shows a small beaker of water placed inside a larger beaker of water. The temperature of the water in beaker A is initially at 90 °C and the temperature of water in beaker B is at 19 °C. Explain what will happen in time.

A02

B-A*

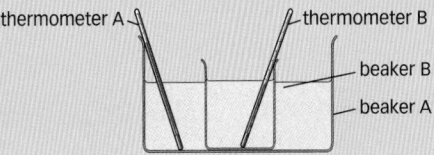

thermometer A — thermometer B — beaker B — beaker A

A01	Recall the science
A02	Apply your knowledge
A03	Evaluate and analyse the evidence

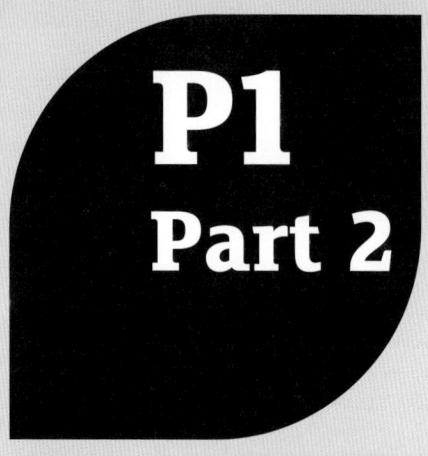

P1 Part 2

Electrical energy and waves

Why study this unit?

Human activity is leading to changes in climate, and we are rapidly using up our natural resources. Scientists are working on solutions to these problems. We consume vast amounts of electricity every day, powering a range of appliances including our TVs and computers. In the future, how will we generate enough to meet our needs?

In this unit you will learn about how electricity is generated, and the advantages and disadvantages of the different technologies, from large coal-fired power stations to small solar cells on calculators.

You will also learn more about waves. How, when we talk, our mobile phones convert sound waves into microwaves before beaming them at high speed to the nearest mobile phone mast. Finally, you will learn about how electromagnetic waves provide the evidence for the Big Bang theory. This is one of the most important of scientific ideas; it describes how everything around us, the entire Universe, was formed.

You should remember

1. All human activity has an impact on the environment.
2. The conservation of energy states that energy cannot be created or destroyed.
3. Electricity is generated in different types of power station.
4. Waves, like light and sound, transfer energy from one place to another.
5. Light and sound can be reflected off objects, and refracted when they travel from one material to another.

The world's largest power station is the Three Gorges Dam on the Yangtze River in China. The project has been in development since 1994. When it reaches full capacity it is expected to be able to produce as much as 22 500 000 000 watts of power (22.5 gigawatts, GW). That's enough electricity for every person in the UK and Australia to watch their own large plasma TV at the same time.

The world's smallest 'power station' is a phytoplankton – a single-celled aquatic organism that converts sunlight into chemical energy to create living biomass.

Learning objectives

After studying this topic, you should be able to:

- ✔ explain that, in some power stations, fuel is used to heat water to produce steam
- ✔ describe how steam drives a turbine connected to a generator
- ✔ describe how naturally occurring steam can be used to drive turbines

Generating electricity

Electricity is generated using sources of energy. A great deal of the electricity in the UK is generated by using an energy source to heat water. The energy source can be **fossil fuels**, biomass or the Sun. This type of power station is called a **thermal power station**.

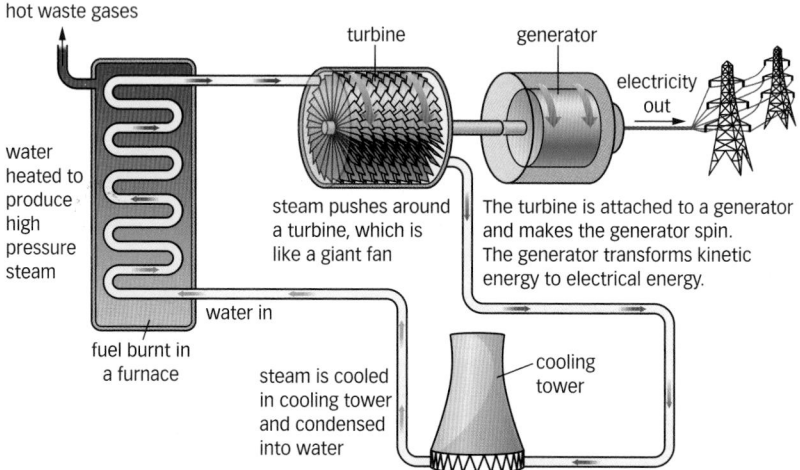

hot waste gases

turbine

generator

electricity out

water heated to produce high pressure steam

steam pushes around a turbine, which is like a giant fan

The turbine is attached to a generator and makes the generator spin. The generator transforms kinetic energy to electrical energy.

water in

fuel burnt in a furnace

steam is cooled in cooling tower and condensed into water

cooling tower

▲ A conventional power station where an energy source is used to heat water

▲ A cooling tower at a power station

Did you know...?

The shape of cooling towers helps to set up a natural convection current. Cool air is drawn in at the base and flows over radiators. Heat from the steam is transferred to the air.

A Why is electricity not a source of energy?

▲ A turbine in a power station being repaired. The turbine is about 5 m across

Geothermal power

In some parts of the world, such as Iceland, where there are volcanic areas, **geothermal** energy can be used to **generate** electricity.

Steam rises to the surface of the Earth, or is not far below it. The steam can be collected and piped to a power station to drive the **turbines** directly.

Geothermal energy does not have any fuel costs, but money is spent on building the power station and maintaining it. These power stations can also be started up and stopped relatively easily.

Geothermal energy can also be used to heat houses. In Iceland, the waste steam from a geothermal power station is piped to houses to heat them. Boreholes can also be drilled into the ground to collect the steam.

Key words

fossil fuel, thermal power station, geothermal power, generator, turbine

▲ In New Zealand, boreholes are drilled into the ground to collect the steam

B What is geothermal energy?

Exam tip **AQA**

✓ Remember that electricity is not a source of energy, but is a form of energy.

Questions

1 What are the differences between a conventional power station and a geothermal power station? ↓ E

2 What does the turbine do?

3 What does the generator do? ↓ C

4 What is the function of the cooling towers?

5 Draw a flow diagram to show all of the energy transfers taking place in a conventional power station. ↓ A*

17: Fossil fuels and carbon capture

▲ Coal is a fossil fuel

Did you know...?

Fossil fuels are a concentrated store of chemical energy. One kilogram of coal stores a lot more energy than one kilogram of wood. So when the fuels are burnt in a power plant they can generate very large amounts of electricity. In Yorkshire, the Drax coal-fired power station produces 7% of the UK's electricity needs, supplying over 2 million homes. You would need around 4000 of the largest wind turbines to generate the same amount of electricity. (There are only around 3000 wind turbines currently in use in the UK.)

Different fossil fuels

Coal, crude oil and natural gas are examples of **non-renewable** energy resources. All three are formed in similar ways. They are the remains of living organisms which died millions of years ago; this is why they are called fossil fuels. Their remains have been squashed and heated in layers of the Earth's crust. This takes a long time, and we are using up fossil fuels far more quickly than they are being formed. They will eventually run out.

All three fuels must be mined and transported, and this can be bad for the environment. However, using fossil fuels for generating electricity has some advantages. The power plants use well established and reliable technology and produce large amounts of electricity. In 2009, around 77% of the UK's electricity came from fossil fuels (mainly coal and natural gas).

> A Give two advantages of using fossil fuels to generate electricity.

In every fossil fuel power station, fuel is burnt and the heat produced turns water to steam. This steam turns a turbine. (Sometimes air is heated directly in gas-fired stations.)

Burning the fuel releases pollutants into the atmosphere.

▲ Burning any fossil fuel releases carbon dioxide into the atmosphere

One of the gases released is carbon dioxide. This contributes to global warming. The different fuels produce different amounts of carbon dioxide, as shown in the table below.

The table shows other differences too. For example, gas power plants have a short start up time. They can very quickly increase the amount of electricity they generate, to cope with sudden increases in demand.

Fossil fuel	Substance heated	Carbon dioxide production	Start up time	% UK electricity supply
coal	water	very high	long	31
oil	water	high	long	1
natural gas	water or air	medium	short	46

Carbon capture

One idea for reducing the amount of carbon dioxide that burning fossil fuels releases into the atmosphere is **carbon capture**. The carbon dioxide is trapped and stored before it enters the atmosphere. One suggestion is that carbon dioxide should be stored in old oil and gas fields such as those found under the North Sea.

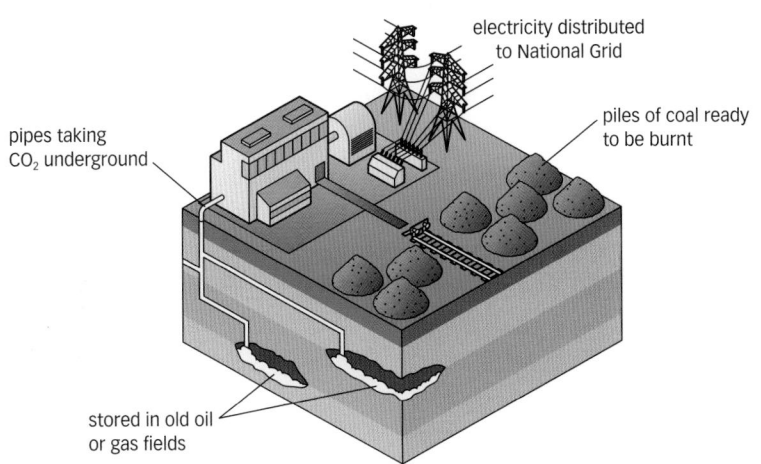

electricity distributed to National Grid

pipes taking CO_2 underground

piles of coal ready to be burnt

stored in old oil or gas fields

▲ Carbon capture will reduce the amount of carbon dioxide that enters the atmosphere

Some scientists have calculated that carbon capture could reduce the amount of carbon dioxide entering the atmosphere from a coal power plant by over 80%. However, this is a rapidly evolving technology and there are currently no carbon capture power stations in the UK.

Key words
non-renewable, carbon capture

B Which fossil fuel produces the least amount of carbon dioxide when burnt in a power plant?

Questions

1 List three fossil fuels and state their percentage contribution to the UK's electricity supply.

2 Explain why fossil fuels are a non-renewable energy resource.

↓ E

3 Explain the process of carbon capture and give an example of where carbon dioxide might be stored.

4 List some of the disadvantages of using fossil fuels to generate electricity.

↓ C

5 Why is it important that natural gas power plants have a short start up time?

↓ A*

18: Nuclear power

Learning objectives

After studying this topic, you should be able to:

- ✔ describe how a nuclear reactor works
- ✔ outline some of the benefits and hazards of using nuclear power to generate electricity

Key words

nuclear reactor, nuclear fission, decommissioning, radioactive waste

A Name two fuels used inside a nuclear reactor.

Inside a nuclear reactor

Nuclear power can be used to generate electricity. Inside a **nuclear reactor,** atoms of uranium or plutonium undergo **nuclear fission**. This releases a huge amount of energy in the form of heat. The heat is used to turn water into steam that drives turbines, as in thermal power plants. No burning is involved, so there is no release of carbon dioxide.

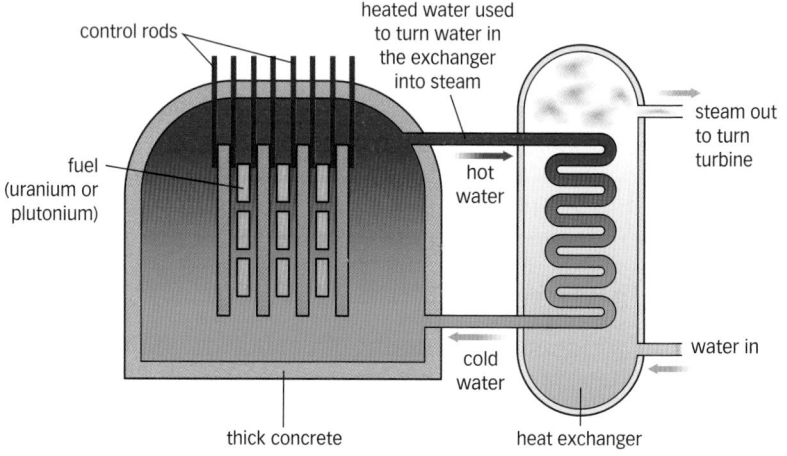

▲ Controlled nuclear fission inside the nuclear reactor releases a huge amount of energy

▲ Uranium is made into pellets that are inserted into a nuclear reactor

Did you know...?

To produce the same amount of electricity you get from 1 kg of uranium, you would need over 15 000 kg of coal!

Arguments for and against nuclear power

Nuclear power is a controversial method of generating electricity. Currently around 13% of the electricity generated in the UK comes from nuclear power and there are plans to build several more nuclear reactors.

Nuclear power has some advantages when compared with other methods:

- Huge amounts of electricity can be generated for each kilogram of fuel used.
- No carbon dioxide is produced, so there is no contribution to global warming.
- The fuel is readily available and won't run out for thousands of years.

There are also disadvantages:

- Nuclear reactors produce highly radioactive nuclear waste. This remains dangerous for millions of years, and so has to be buried deep underground.
- The costs of building the plant and of taking it down when it has finished (called **decommissioning**) can be quite high. This means the electricity generated can be relatively expensive.
- Nuclear reactors have a very slow start up time. It takes a long time to increase or decrease the amount of electricity they are generating.
- There is always the risk of an accident that could release **radioactive waste** into the environment.

▲ Highly radioactive waste is stored in a pond until it can be processed

B How much of the UK's current electricity supply is generated by nuclear power?

Questions

1 List the advantages and disadvantages of using nuclear power.

2 Describe how nuclear reactors can be used to generate electricity.

3 State the energy changes inside a nuclear power plant (ending with electrical energy from the generator).

4 Is nuclear power a renewable or non-renewable energy resource? Explain your answer.

Exam tip AQA

✓ Remember, one advantage of nuclear power is that a very large amount of electricity is generated per kilogram of fuel. It is not enough just to say that a lot of electricity can be generated, as that is also true for fossil fuels.

Key words

biomass, renewable, biofuel

Biomass and biofuels

Biomass is a **renewable** energy resource. There are lots of different types, but all forms of biomass involve material produced by living organisms. Biomass used for burning is called a **biofuel**. These are similar to fossil fuels, but with fossil fuels the living organisms died millions of years ago.

▲ This biomass is ready for burning

Biofuels can be solids, liquids or gases. Some examples are:
- Wood and woodchips, from specially grown trees (new trees are planted when old ones are cut down).
- Alcohol fuels (such as ethanol), produced by fermenting sugar cane crops.
- Methane gas, given off by animal waste in storage tanks called sludge digesters, and also from other rotting waste (for example food waste from homes).
- Nutshells, that are a waste product from manufacturing cattle feed or other food.
- Vegetable oils, that can also be used to make biodiesel (for example, oil from oilseed rape crops). Biodiesel can also fuel cars, buses and even trains.
- Other crops, such as straw.

A Give three examples of different biofuels.

B Give an example of a use for biodiesel other than for generating electricity.

208

In the UK, biomass is used to generate more electricity than any other form of renewable energy. Over 40% of the electricity from renewable sources comes from biomass. The table below shows its advantages and disadvantages compared with other methods of generating electricity.

Advantages	Disadvantages
Uses products which might otherwise be wasted, so the fuel costs are very low	Releases atmospheric pollutants
Power stations can also supply hot water to local industry/houses	In developing countries, land which could be used for food is now used to grow crops for biofuels, leading to food shortages
Carbon neutral	Power plants can be ugly to look at (visual pollution)

Carbon neutral

In a biomass power plant the biofuel is burnt and electricity is generated in a similar way to that in other thermal power stations (water to steam, etc). This releases carbon dioxide into the atmosphere. However, unlike fossil fuels, there is no overall increase in carbon dioxide as the amount released is the same as the plant absorbed while it was alive (as part of photosynthesis). This means biomass is considered to be carbon neutral.

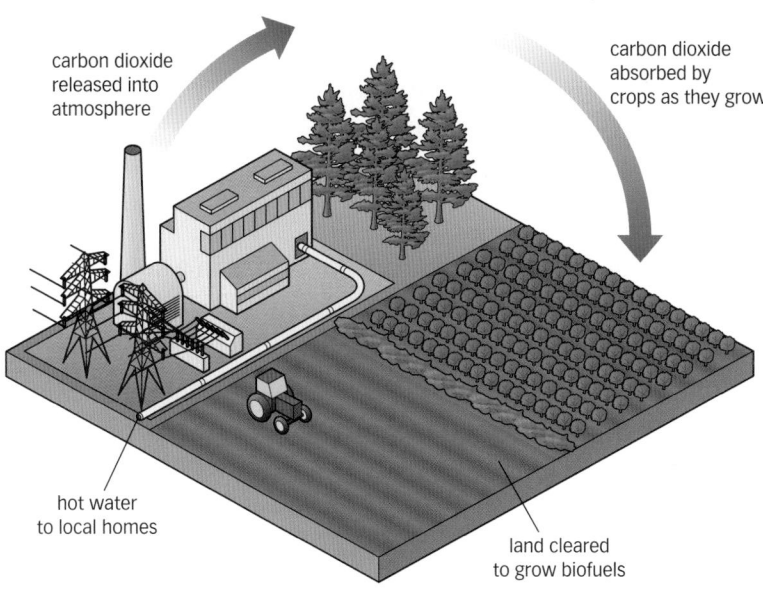

carbon dioxide released into atmosphere

carbon dioxide absorbed by crops as they grow

hot water to local homes

land cleared to grow biofuels

▲ Using biomass is carbon neutral because there is no overall increase in carbon dioxide

▲ Large areas of forest are being cut down to provide land to grow biofuels

Did you know...?

A large chicken and turkey farm in Norfolk uses bird waste to generate all its electricity. In fact, the birds produce so much of it that there is enough electricity left over to power another 5000 homes in the surrounding area. That's a lot of chicken poo!

Questions

1 What percentage of the UK's electricity that is generated from renewable resources comes from biomass?

2 State two advantages and two disadvantages of using biofuels to generate electricity.

↓ E

3 Explain why biomass is described as carbon neutral.

↓ C

4 Describe the similarities and differences between biofuels and fossil fuels.

↓ A*

Learning objectives

After studying this topic, you should be able to:

✔ describe how electricity can be generated using solar power

✔ describe how electricity can be generated using wind turbines

▲ The mirrors reflect sunlight to the top of the solar thermal tower

▲ This solar stove works in the same way as a solar thermal tower. Light from the Sun is reflected on to the cooking pot.

Energy from the Sun

Energy from the Sun can be used to generate electricity as well as heating water. **Solar cells** generate direct current when light energy is absorbed by them. They transfer light energy into electrical energy and produce direct current. The amount of electricity produced depends on the area of solar cell that has light shining on it.

Solar cells do not need much maintenance and there is no need for fuel. There is no need for power cables, so they can be used in remote locations. They do not produce any waste. They are a renewable energy resource. The disadvantages of solar cells are that a large area is needed to generate the same amount of electricity as from a thermal power station, and that less electricity is generated on cloudy days or at night.

▲ A solar cell power station

Another way of using energy from the Sun is in a **solar thermal tower**. A large number of mirrors are used to reflect light from the Sun into one spot at the top of a tower. The temperature can reach 500 °C. The energy is used to heat water into steam. The steam is then used to drive turbines in the same way as in a thermal power station.

> **A** What energy transfer takes place in a solar cell?

Energy from the wind

Winds are convection currents set up in the Earth's atmosphere by energy from the Sun. The kinetic energy of the wind can be used to drive **wind turbines** directly.

Wind turbines are a renewable energy resource. They do not produce any waste, but many people think that they spoil the landscape, and they are also noisy.

Just like solar cells, they can be used in remote locations, but depend upon the wind blowing. They must also be built in open windy areas. A wind farm that could generate as much electricity as a thermal power station would take up a large area.

> **B** What energy transfer takes place in a wind turbine?

How a solar cell produces electricity

When infrared radiation is absorbed by a solar cell, electrons are knocked out of the atoms in the solar cell. These electrons are then free to flow as a direct current.

The amount of electricity produced by a solar cell also depends on the **intensity** of the light shining on it.

▲ Wind turbines

Key words

solar cell, solar thermal tower, wind turbine, intensity

Exam tip

✔ Remember the difference between a solar panel (spread P1.9) and a solar cell. A solar panel transfers energy to water, and a solar cell generates electricity directly.

✔ A solar cell cannot generate electricity at night.

Questions

1 What is a renewable energy resource?

2 What are the differences between a solar thermal tower and a thermal power station?

3 Draw up a table to summarise the advantages and disadvantages of each type of renewable energy described on these pages.

4 Solar cells are used to provide electricity for garden lights. What is needed so that the lights can come on at night?

E

↓ C

↓ A*

Learning objectives

After studying this topic, you should be able to:

✓ explain how energy from waves, tides, and falling water can be used to drive turbines directly

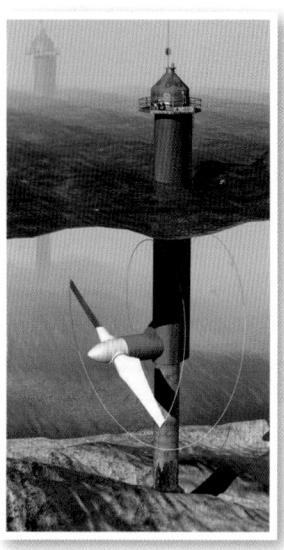

◀ Tidal stream turbine in Strangford Lough, Northern Ireland

A How can electricity be generated from tides?

▲ Generating electricity from waves on Islay, Scotland

Tidal and wave power

Tides cause the sea level to move up and down. When the tide is high, the water can be held back by a **barrage** across a river estuary. After a few hours, the sea level on the other side of the barrage has fallen. Water is allowed to flow through turbines in the barrage to generate electricity.

Tidal barrages are a renewable energy resource – no fuel is needed. However, there are not many places where they can be built, and they can only generate power for 6–8 hours out of every day. Also, the time of high tide is not the same every day – power will be generated in the middle of the night.

▲ The Rance tidal barrage in Brittany, France

Tides also cause tidal streams where seawater is moving from one place to another. In some places the tidal streams are very strong. Moving seawater has kinetic energy and it can be used to drive turbines directly. A **tidal stream turbine** in Strangford Lough in Northern Ireland generates enough electricity to supply around 2500 homes (1.2 MW). It works in the same way as a wind turbine. Tidal streams can be used to generate electricity for 18–20 hours every day. There are plans to install many more tidal stream turbines off the coast of Anglesey in North Wales.

Waves can be used to generate electricity, for example by forcing air up and down a tube. The moving air drives a turbine. The picture on the left shows a wave generator installation on the island of Islay in Scotland.

Hydroelectric power

A **hydroelectric** power station also transforms the kinetic energy in moving water into electrical energy. A dam is built in a hilly area to store water in a reservoir. Water then flows downhill in pipes to the power station. The flowing water drives turbines to generate electricity.

Hydroelectric power stations are a renewable energy resource. They do not need fuel. They can be started and stopped very quickly. However, a reservoir must be built, which can change the environment.

Small hydroelectric power stations can be used in remote areas with high rainfall.

Key words

barrage, tidal stream turbine, hydroelectric

Exam tip **AQA**

✔ Remember the advantages and disadvantages of all the methods of generating electricity – you may need to know them in the exam.

B What is hydroelectric power?

▲ Turbine from a hydroelectric power station

Questions

1 Is fuel needed to generate electricity from moving water?

2 Describe all the energy transfers that take place in a hydroelectric power station.

3 Draw up a table to summarise the advantages and disadvantages of all the renewable energy resources shown on these two pages.

4 How are all these methods of generating electricity similar to what happens in a thermal power station?

5 Ellen says 'waves can always be used to generate electricity'. Is she correct? Explain your answer.

E

C

A*

Learning objectives

After studying this topic, you should be able to:

- explain how electricity is transferred to consumers
- explain how transformers are used in the National Grid to minimise losses when energy is transmitted

▲ An electricity substation

Distributing electricity

When electricity has been generated it needs to be distributed to consumers in homes, shops and offices. The electricity you use is transferred around the country by the **National Grid**. This is the **mains supply** that is available whenever you switch on an appliance at home.

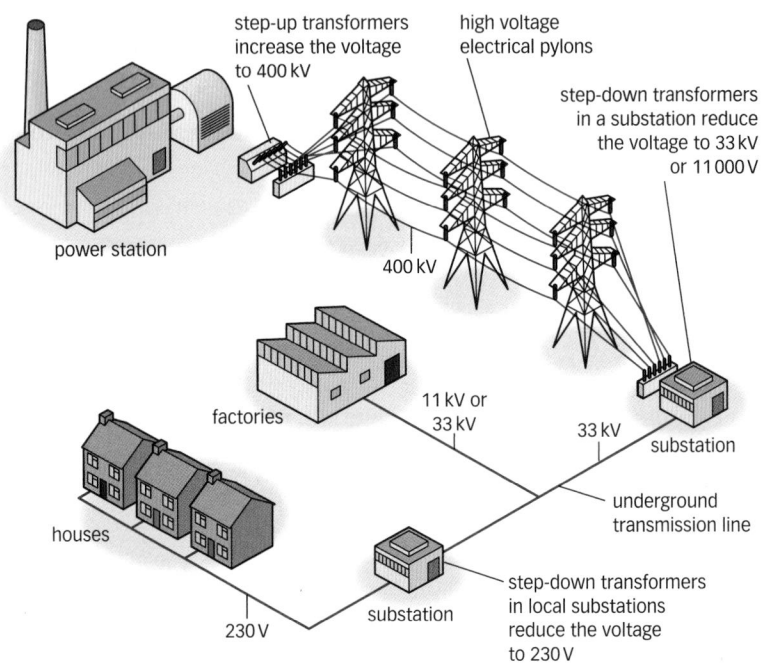

step-up transformers increase the voltage to 400 kV

high voltage electrical pylons

step-down transformers in a substation reduce the voltage to 33 kV or 11 000 V

power station

400 kV

factories

11 kV or 33 kV

33 kV

substation

houses

underground transmission line

230 V

substation

step-down transformers in local substations reduce the voltage to 230 V

▲ The National Grid

Transformers

The electricity goes from the power station to a **step-up transformer** where the **voltage** is increased. Electricity is transferred from the transformer by high-voltage transmission lines to an electricity **substation**. Here, the electricity goes through a **step-down transformer** to reduce the voltage to a safer level for use by consumers.

A step-up transformer increases the voltage but reduces the current. This minimises energy loss in transmission.

A What is the difference between a step-up and a step-down transformer?

B What happens in a substation?

Power lines

Large amounts of energy are transferred by the National Grid. The higher the current in the transmission cables, the greater the energy losses during transmission.

When a current flows through a wire, there is a heating effect and this dissipates some energy. The larger the current, the more heat is produced. This means that less energy is transferred when the current is higher.

Some power lines are carried overhead. Others are buried underground, so they cannot be seen and there is not the danger of tall vehicles or other objects tangling with them. However, it is much more expensive to bury power lines underground. They have to be insulated and waterproofed. Also, if they have to be repaired or renewed, the ground must be dug up.

▲ Repairing a power line

Key words

National Grid, mains supply, step-up transformer, voltage, substation, step-down transformer

Exam tip

✔ Remember all the parts of the National Grid – you may need to name them in the exam.

Questions

1 What is the National Grid? ▼ E

2 Explain how electricity is transferred from a power station to your home through the National Grid.

3 (a) What are the advantages of putting cables underground to distribute electricity?

   (b) What are the disadvantages? ↓ C

4 Why is electricity stepped down before it enters the home?

5 Why is electricity stepped up before being distributed by the National Grid? ↓ A*

Learning objectives

After studying this topic, you should be able to:

✔ evaluate ways of matching supply with demand

Key words

pickup, start up time, pumped storage

Demand for electricity

The demand for electricity varies according to the time of year and the time of day. The demand can also change from minute to minute, depending on what is happening. If many people are watching a TV programme, the demand for electricity falls. When the programme ends, demand for electricity can suddenly increase because people turn on kettles or lights. This is called a **pickup**.

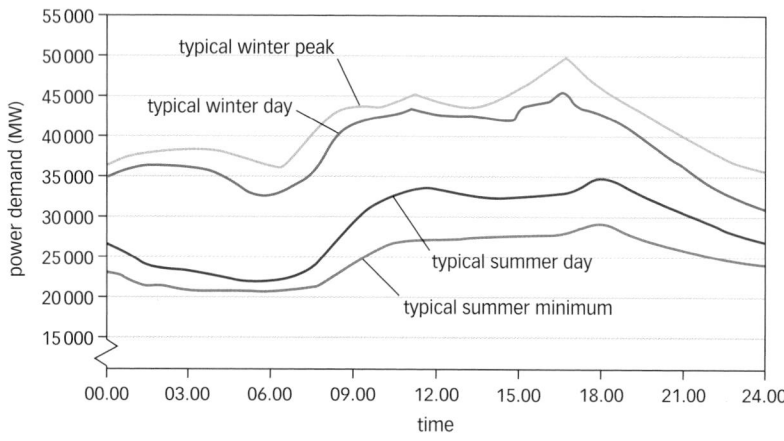

▲ Typical electricity demand for different days in the year

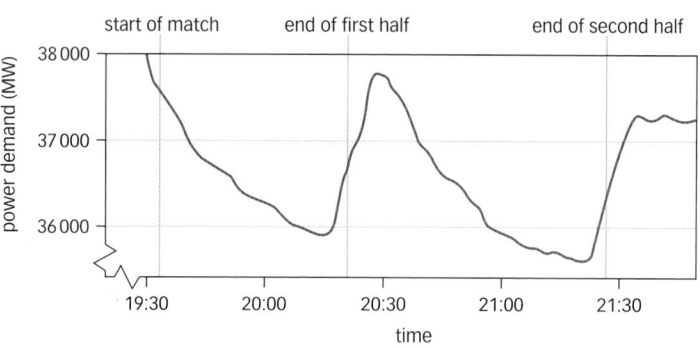

▲ How the electricity demand can vary when a football match is shown on TV

Did you know...?

The biggest TV pickup in the UK was in 1990 when England played Germany in the football World Cup semi-finals. The game went to a penalty shoot-out. When the programme finished, demand increased by 2800 MW.

A What is a pickup in electricity demand?

B Why are hydroelectric power stations used to generate electricity during a pickup?

It is wasteful to keep generating electricity at a high level. So power engineers stop some power stations from generating electricity when there is a low demand. They start them up again when there is a high demand.

Power engineers have to forecast when there is likely to be a high demand, so that there is the right amount of power available. They look at weather forecasts and TV schedules.

They keep some power stations generating electricity, even though not all of the electricity is needed. But they can also use hydroelectric power stations, because they can be switched on very quickly. It only takes a couple of minutes to start up a hydroelectric power station as they have a very short **start up time**.

It can take hours to start and stop fossil fuel power stations, so they are often left running. The extra electricity they generate is used in **pumped storage** hydroelectric power stations. At times of low demand, energy is stored at a hydroelectric station by pumping water from a lower reservoir to a higher one as shown in the diagram.

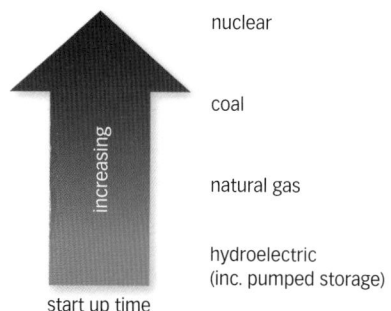

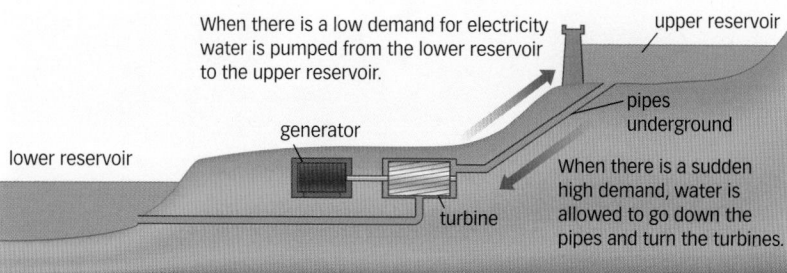

▲ A pumped storage power station

▲ Start up times vary for different power stations

▲ Dinorwig in North Wales is an example of a pumped storage station found in the UK

Questions

1 What factors do power engineers have to think about when forecasting what the demand for electricity might be?

2 More people are using air-conditioning. How do you think this will affect the electricity demand curves?

3 Draw a flow chart to show the energy transfers happening in a pumped storage power station when:
   (a) it is pumping water from the lower reservoir to the higher reservoir
   (b) it is generating electricity.

4 When do pumped storage power stations pump water to the higher reservoir?

5 What events other than a football match could generate a big pickup in the demand for electricity?

Exam tip **AQA**

✔ Remember that a pumped storage power station is a hydroelectric power station where water can also be pumped back up to the higher reservoir.

Learning objectives

After studying this topic, you should be able to:

- describe how waves transfer energy from one place to another
- explain the terms frequency, wavelength and amplitude
- use the equation $v = f\lambda$

Key words

oscillation, vibration, energy, amplitude, wavelength, frequency, wave equation

What are waves?

Waves are around us all of the time. We see water waves on the surface of the sea, sound waves allow us to listen to our music – and just think what the world would be like without light.

▲ Surfing some water waves

All waves are a series of **oscillations** (or **vibrations**) which travel from one place to another. Water waves make water molecules move up and down, and sound waves make air particles vibrate from side to side. In all cases the waves transfer **energy** from one place to another.

▲ Energy is transferred from the speaker to your ears

Did you know...?

Some waves can have very high frequencies measured in kHz (kilohertz; 1 kHz = 1000 Hz) and MHz (megahertz; 1 MHz = 1 000 000 Hz). An FM radio station might transmit waves with a frequency of 97.7 MHz – that's 97.7 million waves every second!

A What do waves transfer from one place to another?

All waves have some key features.

Amplitude in metres (m)	Maximum height of the wave measured from the middle.
Wavelength in metres (m)	Shortest distance between a point on a wave and the same point on the next wave. For example, the distance from one peak to the next peak.
Frequency in hertz (Hz)	The number of waves passing a point per second. A frequency of 6 Hz would mean six waves pass a point every second.

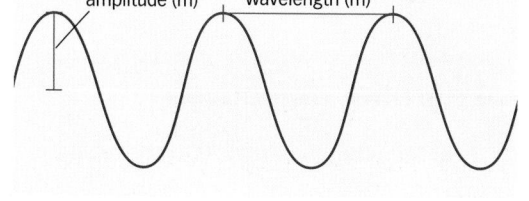

▲ Key features of a wave

The wave equation

The speed of a wave depends on its frequency and its wavelength. They are related in the **wave equation**:

$$\begin{matrix} \text{wave speed} \\ \text{(metres/second, m/s)} \end{matrix} = \begin{matrix} \text{frequency} \\ \text{(hertz, Hz)} \end{matrix} \times \begin{matrix} \text{wavelength} \\ \text{(metres, m)} \end{matrix}$$

If the wave speed is called v, the frequency f and the wavelength λ (a greek letter, pronounced lam-da), then $v = f\lambda$

Worked example 1

Dylan is standing on the end of a pier. He measures the water waves going past him. The wavelength of each wave is 1.3 m. He counts 2 waves every second. Find the wave speed.

wave speed = frequency × wavelength

f = 2 Hz (as there are two waves every second), and λ = 1.3 m, so

$v = 2 \times 1.3$

$v = 2.6$ m/s

Worked example 2

A flute produces a note with a wavelength of 75 cm. The speed of sound is 330 m/s. Find the frequency of the note.

wave speed = frequency × wavelength, to:

$$\text{frequency} = \frac{\text{wave speed}}{\text{wavelength}} \text{ or } f = \frac{v}{\lambda}$$

$f = \dfrac{330}{0.75}$

$f = 440$ Hz

Exam tip AQA

✔ Remember to look carefully at the units when using the wave equation. Pay particular attention to wavelength. This must be measured in metres.

Questions

1 Define the terms wavelength, frequency and amplitude.

2 Draw a wave with an amplitude of 3 cm and a wavelength of 8 cm.

3 Find the speed of a wave with a wavelength of 30 m and a frequency of 120 Hz.

4 A speaker produces a sound at a frequency of 6.6 kHz. The wavelength of the sound wave is 5.0 cm. Use these values to show that the speed of sound is 330 m/s.

5 A radio station transmits waves with a frequency of 120 MHz. The radio waves travel at 3×10^8 m/s. Find the wavelength of the radio wave.

Learning objectives

After studying this topic, you should be able to:

✔ describe the differences between transverse and longitudinal waves

✔ give several examples of transverse and longitudinal waves

Key words

transverse wave, longitudinal wave, compression, rarefaction

A Name the two types of wave.

B Give two examples of transverse waves.

Different types of wave

There are two different types of wave, **transverse waves** and **longitudinal waves**. Both types are made up of oscillations or vibrations, and they both transfer energy from one place to another. The oscillations which make up the waves are slightly different.

Transverse waves

If you were asked to sketch a wave, you would probably draw a transverse wave. They look like ripples on a pond, with peaks and troughs. In a transverse wave the oscillations are perpendicular (at right angles) to the direction of energy transfer.

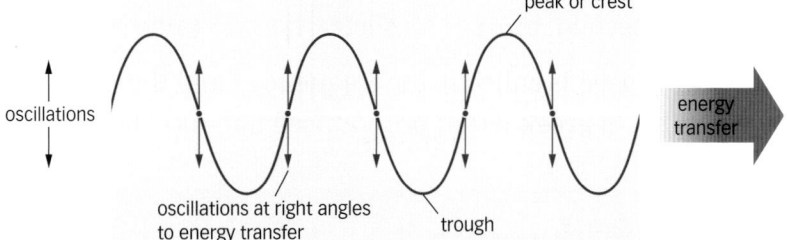

▲ A transverse wave

Examples of transverse waves include:

• water waves
• light
• microwaves
• X-rays
• mechanical waves on strings (such as a plucked guitar string)
• some mechanical waves on springs.

▲ Ripples on the surface of a pond are transverse waves

Longitudinal waves

In longitudinal waves the oscillations are to and fro along the same direction as the energy transfer. They are parallel to the direction of energy transfer.

Sound is an example of a longitudinal wave. If you look closely at a speaker, you can see the speaker cone moving in and out. This creates oscillations which travel through air in the same direction as the speaker movement. When the speaker moves out, it creates a **compression** as the air is bunched up. When it moves back in, it creates a **rarefaction**, where the air is more spread out. It is these compressions and rarefactions which make up a longitudinal wave.

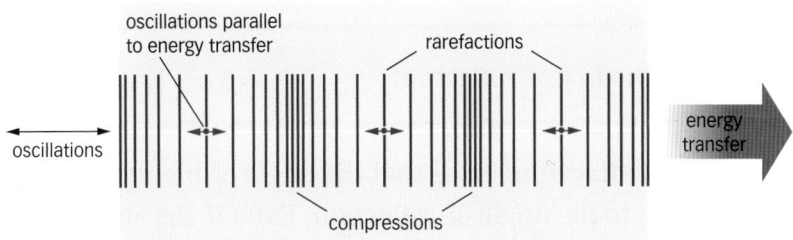

▲ A longitudinal wave

Other examples of longitudinal waves include:
*   one type of seismic wave (in earthquakes)
*   some mechanical waves on springs.

Questions

1   Sketch a diagram showing a transverse wave and one showing a longitudinal wave. Label the key features.

2   Give two examples of longitudinal waves.

3   Describe the differences between transverse and longitudinal waves.

4   Describe how a speaker produces compressions and rarefactions.

5   Describe how a spring could be used to demonstrate both transverse and longitudinal waves.

↓ E

↓ C

↓ A*

Did you know...?

An earthquake produces both transverse (S) and longitudinal (P) waves. These travel at high speed through the Earth. Scientists have carefully recorded when these waves reach the surface. They have used this information to learn more about the structure of the Earth. Thanks to the differences between these types of wave we now know that the Earth has a solid iron inner core around 2000 km across, surrounded by a layer of molten iron 1000 km deep.

▲ A spring can be used to show a longitudinal wave

Exam tip    AQA

✓ When describing a transverse wave, you must say that the vibration is perpendicular, not that the wave vibrates 'up and down' – the movement could be side to side.

Learning objectives

After studying this topic, you should be able to:

- ✔ describe how waves can be reflected, refracted and diffracted
- ✔ use a normal line when drawing ray diagrams

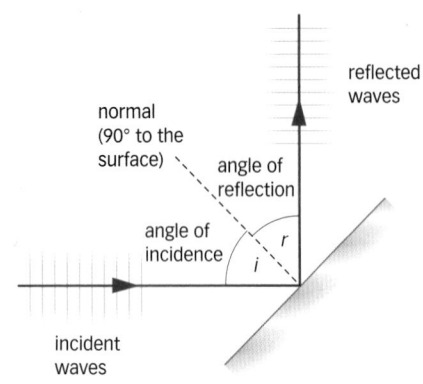

▲ Reflection of a wave

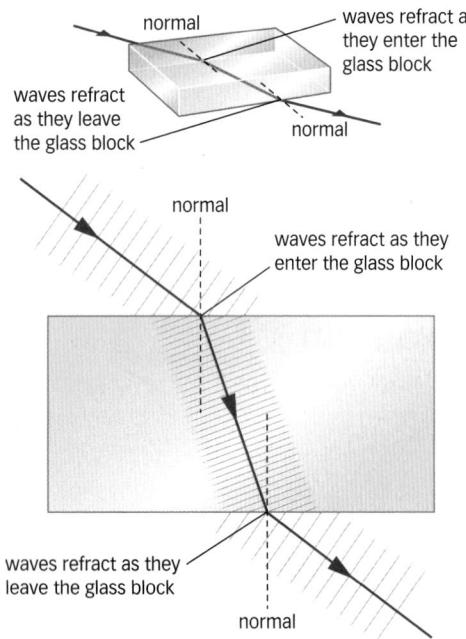

▲ Light refracting as it passes through a glass block

Reflection

When you look in a mirror you see your reflection. But this doesn't only happen with light waves – all waves can be **reflected**. An echo is a reflection of sound waves, and radio waves are often reflected off buildings and interfere with your TV signal.

We can draw simple ray diagrams to show reflection. The rays show the direction of the energy transfer by the waves.

Whenever we draw ray diagrams we must include a **normal** line. This is a line at 90° to the surface. We always measure angles to the normal.

> **A** State the law of reflection.

The **law of reflection** states that the angle of incidence is always equal to the angle of reflection. Even if the surface is really rough, the two angles are always the same.

Refraction

A material that a wave travels through is called a **medium**. When sound travels through air, the medium is air. When waves go from one medium to another they can be **refracted**. As they enter a different medium their speed changes and this causes them to change their direction. If the wave slows down it bends towards the normal, if it speeds up it bends away from the normal.

If the waves travel along the normal then, although their speed changes, they don't change their direction.

> **B** Explain what is meant by the medium that waves travel through.
>
> **C** Describe what happens to a ray of light when it enters a glass block.

Diffraction

Whenever waves pass through a gap or move around an obstacle, they spread out. This is called **diffraction**, and it happens with both transverse and longitudinal waves.

You might have noticed the effect with sound. If you have your door open and someone is talking outside, even though you can't see them, you can hear their voice. The sound waves diffract when they go through your open doorway, spreading out and filling the room. It might sound as though the sound wave actually comes from the doorway itself.

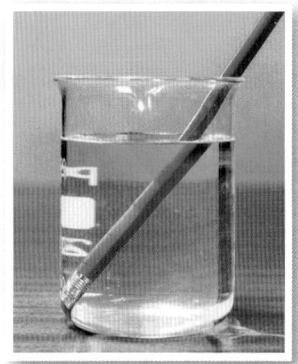

◀ Refraction can lead to some strange optical effects!

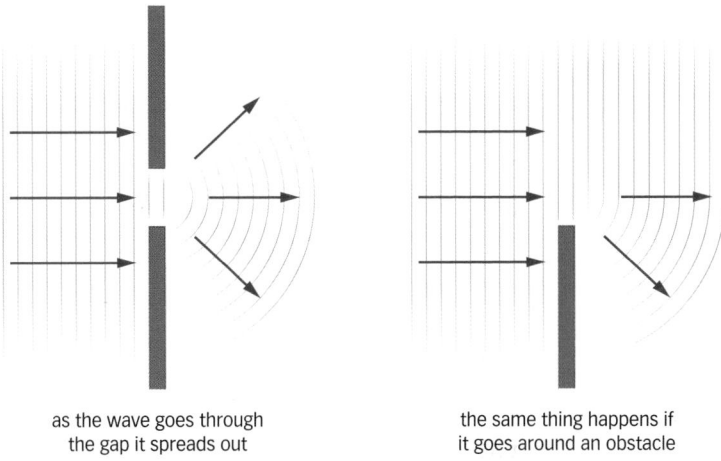

as the wave goes through the gap it spreads out

the same thing happens if it goes around an obstacle

▲ Diffraction through a gap or around an obstacle

Like all waves, the radio waves used for transmitting radio and television signals diffract when passing over obstacles. This allows signals to reach the bottom of some valleys.

The size of the gap and wavelength of the wave affect how much diffraction takes place. The longer the wavelength, the greater the diffraction. The greatest diffraction takes place when the wavelength is around the same size as the gap.

▲ Water waves are diffracted when they enter a harbour. Radio waves may be diffracted by a hill or other obstacle.

Did you know...?

Refraction can make swimming pools appear shallower than they actually are. A pencil can appear bent, and mirages in deserts are caused by refraction.

Key words

reflection, normal, law of reflection, medium, refraction, diffraction

Questions

1 Draw a diagram to show the law of reflection.

2 What types of waves can be diffracted?

E

3 Draw two diagrams to show how waves can be diffracted.

4 Explain why waves are refracted.

C

5 Draw a diagram to show how light is refracted when it travels from water to air.

6 Explain why in some valleys you can't get a TV signal, but you can pick up longer wavelength radio waves.

A*

27: The electromagnetic spectrum

Learning objectives

After studying this topic, you should be able to:

✔ describe the key features of an electromagnetic wave

✔ list the order of waves within the electromagnetic spectrum in terms of wavelength, frequency, and energy

✔ describe some of the hazards and uses of the higher frequency waves in the spectrum

Key words

electromagnetic wave, vacuum, electromagnetic spectrum

B List the electromagnetic waves from longest wavelength to shortest.

Did you know...?

According to Einstein's special theory of relativity, nothing can travel faster than the speed of light through a vacuum. Some science fiction shows on TV bend this rule and use 'warp drives', 'wormholes' or other fanciful future technology. However, the speed of light remains the ultimate speed limit.

Electromagnetic waves

Light, microwaves and X-rays are examples of **electromagnetic waves**. These are a special kind of transverse wave. They do not need a medium such as air or water to travel through. Electromagnetic (EM) waves can travel through a **vacuum** like space. This is how light and infrared waves reach the Earth from the Sun. If electromagnetic waves could not travel through a vacuum, then there would be no way to receive energy from the Sun. Life on Earth would not even exist.

A What makes electromagnetic waves special compared with other transverse waves?

The different electromagnetic waves form a family called the **electromagnetic spectrum**. This is a continuous spectrum with a very wide range of wavelengths. The longest electromagnetic waves are radio waves, which can be over 10,000 m long. Gamma rays have the shortest wavelength, as small as 10^{-15} m. That's a millionth of a billionth of a metre!

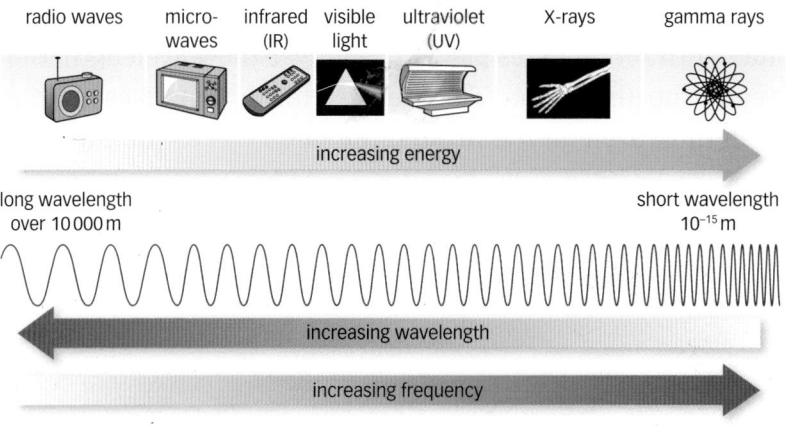

▲ The electromagnetic spectrum

Although they have different frequencies, energies and wavelengths, all electromagnetic waves travel at the same speed through space. This is the speed of light, or 300 000 000 m/s (3×10^8 m/s).

Uses and hazards of electromagnetic waves

The different parts of the electromagnetic spectrum have many different uses, from cooking dinner to communicating with satellites. The properties of ultraviolet, X-rays, and gamma rays make them ideal for a variety of purposes.

Exam tip  AQA

✔ Don't be tempted to say that higher energy waves such as gamma rays, travel faster – it's not true.

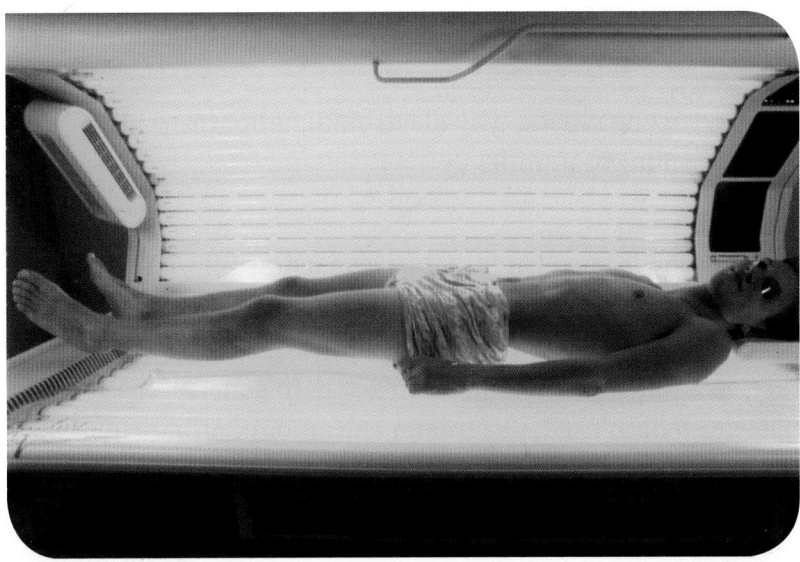

▲ Sunbeds emit ultraviolet, giving you a tan, but they can be hazardous

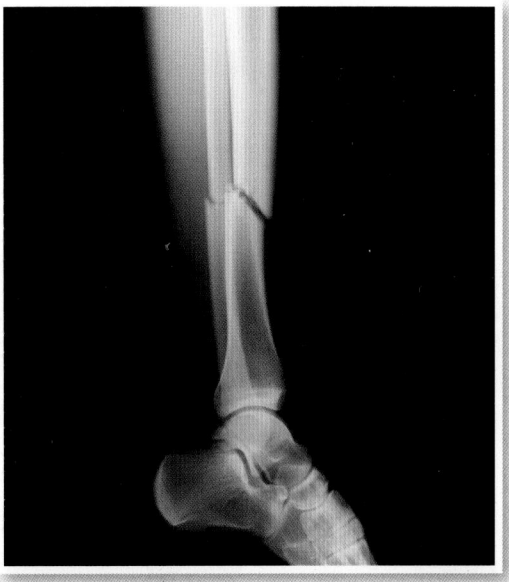

▲ X-ray images help diagnose broken bones

Electromagnetic wave	Some uses
ultraviolet	detecting fake banknotes, sun tanning, sterilising drinking water
X-rays	detecting broken bones, looking for defects in metal products
gamma rays	sterilising medical equipment, killing cancerous cells

There are also hazards associated with the different parts of the electromagnetic spectrum. In particular, the three parts of the spectrum mentioned above can be very hazardous. These electromagnetic waves have very short wavelengths and transfer the most energy. They can travel through many materials and potentially damage or kill human cells.

Questions

1 Give three examples of electromagnetic waves. ↓E

2 List the electromagnetic spectrum from highest frequency to lowest frequency.

3 Waves from which part of the electromagnetic spectrum travel the fastest? Explain your answer. ↓C

4 State two uses for gamma rays.

5 Why are X-rays potentially hazardous and what measures, can be taken to reduce the risks when using them? ↓A*

Learning objectives

After studying this topic, you should be able to:

✔ describe examples of how electromagnetic waves are used to communicate

✔ outline the possible risks involved in using a mobile phone

✔ show how a mirror may be used to produce an image

Using electromagnetic waves to communicate

As electromagnetic waves travel so fast, they are very useful for communications. Every time you listen to the radio, watch TV or chat to a friend on your mobile, you are using electromagnetic waves to communicate.

> **A** Give one reason why electromagnetic waves are used to communicate.

Information is encoded into the wave using different techniques. It is then sent from a transmitter to a receiver (such as a mobile phone or TV aerial). When the wave is received, this information is extracted. In general, the shorter the wavelength used, the greater the amount of information that can be sent per second. The table shows some examples of how we use electromagnetic waves to communicate.

radio waves	TV, radio and wireless communications (like Bluetooth and WiFi)
microwaves	mobile phones and satellite TV
infrared	remote controls and some cable internet connections
light	photography and some cable internet connections

▲ Mobile phones use microwaves to communicate

Did you know...?

Mobile phones transmit and receive microwaves. Research has shown these waves can cause a small heating effect in the brain. Scientists are not sure whether using mobiles from a young age will cause any long-term damage. The current advice from the NHS is that mobiles should not be used regularly by younger children.

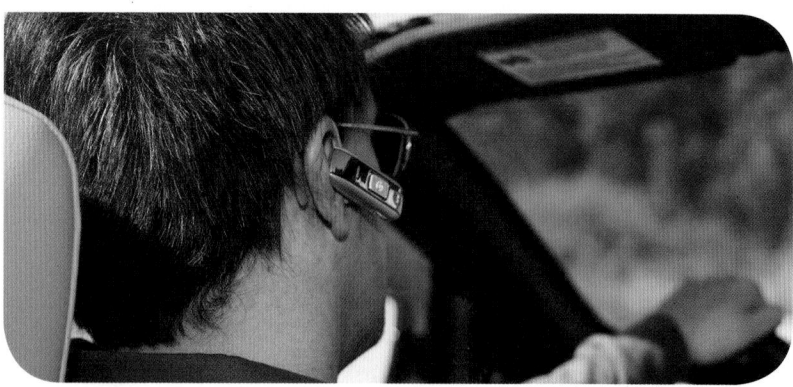

▲ Drivers must use hands-free wireless devices such as Bluetooth headsets if they want to talk and drive at the same time

> **B** Give three examples of electromagnetic waves used in communications.

Images

Visible light only forms a very small part of the electromagnetic spectrum. Our eyes are able to detect this tiny region of electromagnetic waves. This makes light very useful for communicating. Our brain forms images of the world around us when the light reflected off objects enters our eyes.

Our brain also forms an image when light is reflected off a flat, or plane, mirror. But the brain assumes that because light travels in a straight line, the light has come from somewhere in the mirror. This is why mirrors seem to have depth – it looks to us as though there is an image inside the mirror! This image is not really there, and so it is called a **virtual image**.

> C   What is the name of the type of image seen inside a mirror?

Images seen in flat mirrors stay the same way up as the object and the mirror (they are **upright**).

When light reflects off a mirror, there is another effect. You've probably noticed when you look at your reflection it looks as though you have been flipped horizontally. Close your left eye and the reflection in the mirror closes its eye on its right. As a result, the images formed in plane mirrors are described as being **laterally inverted**.

Key words

virtual image, upright, laterally inverted

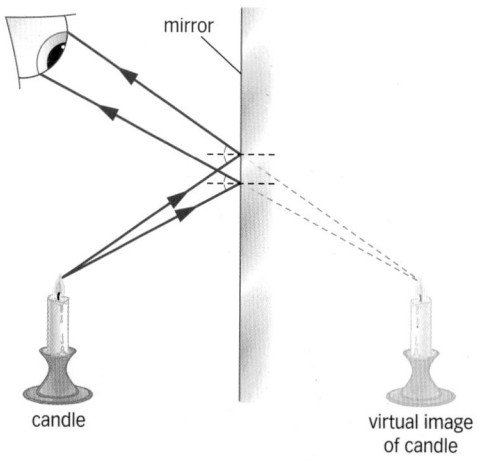

▲ A plane mirror produces an upright, laterally inverted virtual image

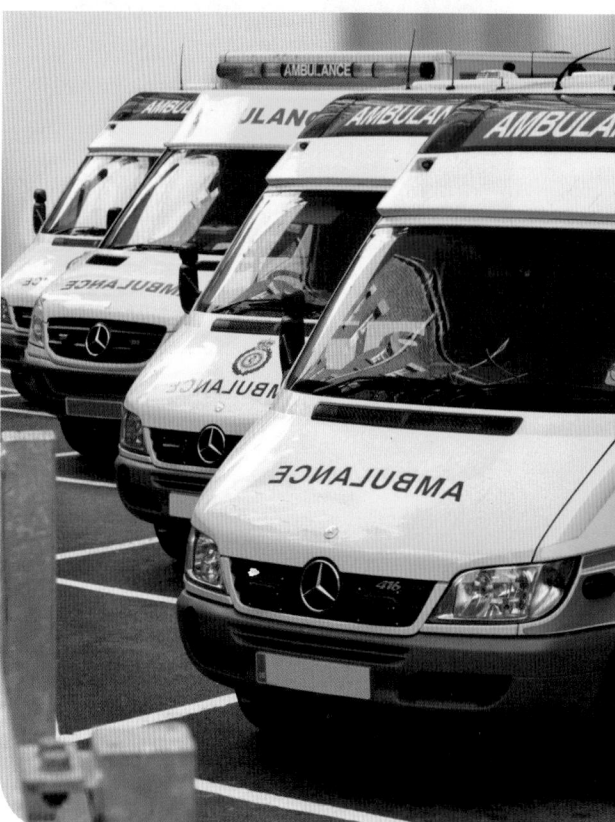

▲ Emergency vehicles have reverse writing on the front, so drivers see it the right way round in their rear-view mirrors

Questions

1   Name the type of electromagnetic wave used by mobile phones.

2   Explain why electromagnetic waves are useful for communication.

3   Carefully draw a diagram to show how an image forms in a mirror.

4   Which three terms describe the image formed by mirrors?

5   Explain why TV signals are generally a shorter wavelength than radio signals.

Learning objectives

After studying this topic, you should be able to:

✔ describe sound waves as longitudinal waves

✔ understand how the pitch of a sound is determined by its frequency, and how the volume is determined by its amplitude

Key words

sound waves, pitch, volume, ultrasound

Did you know...?

As you get older, you lose your ability to hear higher frequency sounds. By the time you are 20 you will probably only be able to hear up to around 16 000 Hz; this falls to 13 000 Hz at 30. But it is not just age that affects your hearing. Listening to music that is too loud, particularly through headphones, does permanent damage.

Listening to sound

Sound waves are created whenever an object vibrates. When you talk, your vocal chords vibrate. You can feel them if you gently press the front of your throat whilst talking. Your headphones contain tiny little speakers which vibrate in and out when they receive an electric current from your MP3 player. If you play a musical instrument, it is often a string, a column of air or a reed that is made to vibrate. These vibrations make the air around the instrument vibrate. These vibrations travel through the air as a sound wave.

▲ As the drumskin vibrates, it creates a sound

When vibrations reach our ear, they make our eardrum vibrate and we detect this as a sound.

> A Describe how sound waves are formed.

Sound can travel through other media too, not just air. Sound can travel through all solids, liquids and gases. The denser the material the sound travels through, the faster the sound. It travels at 330 m/s through air, but 1500 m/s through water and even over 5000 m/s through metals like steel.

Although it can travel through a number of different materials, sound can't travel through a vacuum. There are no particles in the vacuum to vibrate. Just like the tagline of the film says, in space no-one can hear you scream!

> B Other than air, give three examples of materials that sound can travel through.

Pitch and frequency, volume and amplitude

The faster something vibrates, the greater the number of vibrations every second. This means more sound waves are created each second, and so the frequency of the sound wave increases. The higher the frequency of a sound wave, the higher its **pitch**. A high-pitched sound has a high frequency. The loudness of the sound (its **volume**) depends on the amplitude of the sound wave. Louder sounds have much larger amplitudes.

A special piece of laboratory equipment called an oscilloscope can be used to produce a picture of a sound wave. If the sound increases in pitch, the waves get closer together on the screen. Their frequency has increased.

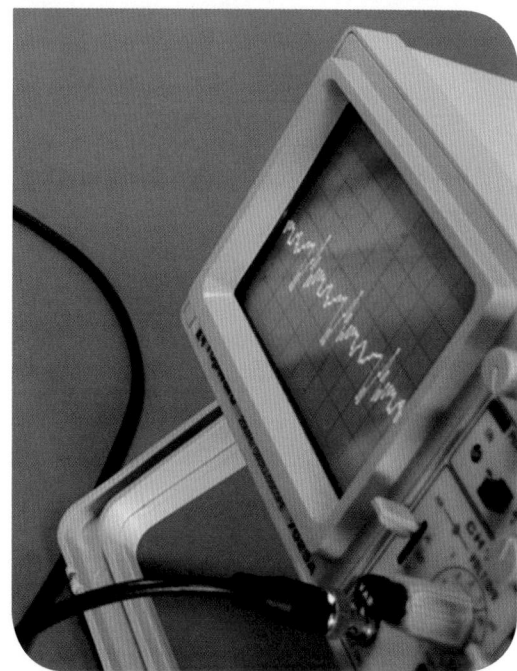

▲ An oscilloscope can produce an image of a sound wave

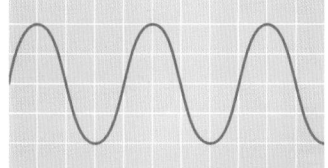

a low pitch, low frequency, loud sound   a higher pitch, higher frequency, quieter sound

▲ Sound waves with different pitch shown on an oscilloscope screen

An oscilloscope produces a picture of a sound wave that is easier to understand. This looks like a transverse wave, but remember that sound waves are longitudinal waves.

Humans can hear sounds with a frequency up to 20 000 Hz. Sound waves above this frequency are called **ultrasound**. Different animals can hear a range of different frequencies of sound. The frequency of a dog whistle is too high-pitched for humans to hear. Dogs can hear much higher frequency sounds, and so the dog hears the whistle and comes running.

▲ The frequency from a dog whistle is too high for humans to hear, but it's fine for dogs

Questions

1. Explain why sound can't travel through a vacuum.

2. Describe how frequency and amplitude affect the pitch and loudness of a sound wave.

3. Describe what happens to the speed of sound as it passes through denser materials.

4. Sketch the oscilloscope traces to compare two waves, one at 200 Hz, the other at 100 Hz.

Exam tip   AQA

✔ We often draw sound waves to look like transverse waves. This makes them easier to draw, but remember that they are actually longitudinal waves.

Learning objectives

After studying this topic, you should be able to:

- ✔ describe the Doppler effect
- ✔ explain why a change in wavelength and frequency is observed when a wave source is moving

Key words

Doppler effect, wave source

Moving wave sources

If you've ever heard an F1 car race past, you will have heard it make that distinctive *neeeeeaaaaawwww* sound. This is an example of the **Doppler effect**. The sound comes from the engine of the car. At top speed, it produces a sound at a constant pitch. However, because the source of the sound (the car engine) is moving, the note sounds different depending on whether the car is moving towards or away from you. There is a change in the pitch of the sound as the car races past.

▲ You can clearly hear the Doppler effect as an F1 car races past

When the car is approaching you, it makes the *neeeee* sound. The pitch sounds higher and higher; the waves have a higher frequency. When the car has gone past, it makes the *aaaawwww* sound. The sound gets lower pitched; the sound waves have a lower frequency.

Anything that emits waves is called a **wave source**. A car engine is a wave source, emitting sound waves. A light bulb is a wave source, emitting light. The Doppler effect happens with all waves, not just sound. Where the wave source is moving towards an observer, there is an increase in frequency. Where it is moving away, the waves have a lower frequency.

The same thing happens if the wave source is stationary and the observer moves towards or away from it. But you have to be moving really fast to notice the effect.

A What is the name given to the effect that causes a change in pitch when an F1 car travels past a stationary observer?

B If the sound has a higher pitch, does the sound wave have a lower or higher frequency?

Why does this happen?

When the source of waves is moving towards the observer, the waves it emits are all bunched up. They are compressed together, and this gives them a shorter wavelength. As a result there are more waves per second and so the wave has a higher frequency.

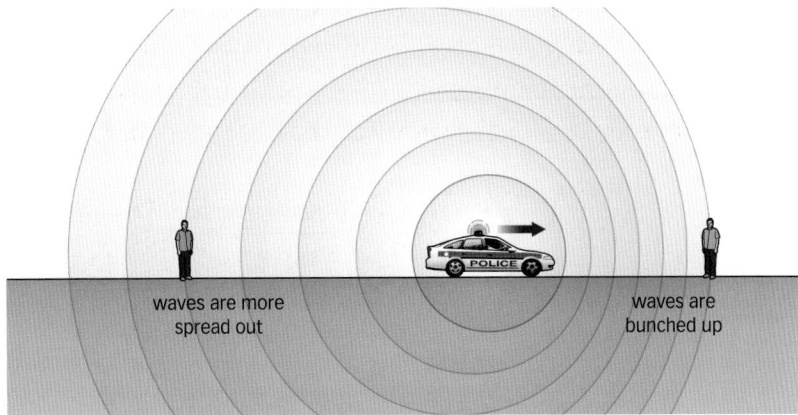

waves are more spread out

waves are bunched up

▲ The Doppler effect can be heard when a police car passes you with its siren sounding

When the wave source is moving away, the wavelength is stretched, the waves are much more spread out. Fewer waves arrive at the observer each second, and so the frequency is lower.

▲ The Doppler effect is used to measure vehicle speeds

Did you know...?

The Doppler effect can be used to calculate an object's speed. Radar guns are used to measure the speed of many things, including the speed of a tennis ball after a serve and the speed of a football struck into the back of the net. They are also found in speed cameras used to monitor vehicle speeds. The guns emit radio waves which reflect off the vehicle being measured. If the vehicle is moving towards the camera, the reflected waves are more bunched up than when they were fired. Their wavelength has gone down due to the Doppler effect. The faster the vehicle is moving, the greater the change in wavelength, and so its speed can be determined very accurately.

Questions

1 Explain what is meant by wave source, and give two examples.

E

2 Describe what causes the Doppler effect

C

3 Explain what would happen to the change in pitch if the F1 car was moving much faster.

4 Describe what the driver hears as they race along a straight. Explain your answer.

A*

Learning objectives

After studying this topic, you should be able to:

✔ know the meaning of the term red-shift

✔ describe how red-shift is observed in light from distant galaxies

Key words

red-shift

What is red-shift?

The Doppler effect happens to all waves, not just sound. The police rely on the Doppler effect from radio waves or microwaves to detect speeding motorists. Even wavelength of light changes if the light source moves away from, or towards you.

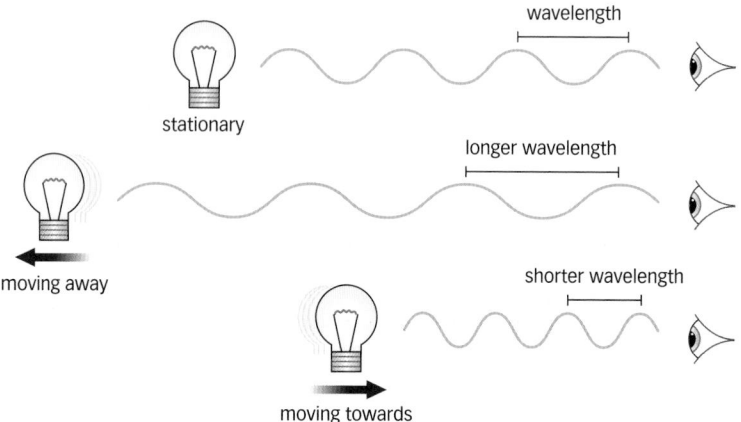

▲ Light wavelength shifting

If a light source moves away from you, the light from it gets stretched. Its wavelength increases, the waves get longer. Red light has the longest wavelength of any colour, so this effect is called **red-shift**. The light is shifted towards the red end of the visible spectrum.

The opposite happens if the light sources moves towards you.

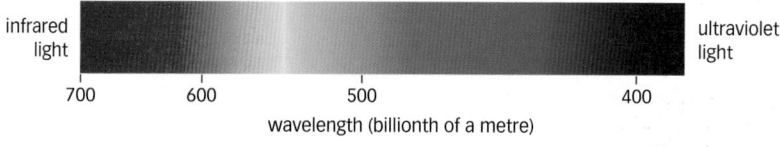

▲ Visible spectrum

Unlike sound, you don't notice the change. Light travels too fast; an object would have to be moving at close to the speed of light before it looked any different to the naked eye. It would look really odd if a car's headlights looked blue when it came towards you, but red when it moved away!

A What colour in the visible spectrum has the longest wavelength?

Galaxies and red-shift

It turns out when you look at the light from most of the other galaxies in the Universe, it is red-shifted. This must mean most galaxies are moving away from us.

▲ A galaxy

> **B** Are most galaxies moving towards or away from us?

When scientists carefully analysed this light they noticed another very important pattern. The galaxies furthest away from us showed a greater red-shift. These galaxies must be moving faster.

These observations on red shift lead to the conclusion that the further away a galaxy is from us, the faster it is moving.

Did you know...?

It was the American astronomer Edwin Hubble who first explained the red-shift in distant galaxies. The fact that more distant galaxies are moving faster is called Hubble's law. For such an important contribution to astrophysics, the Hubble Space Telescope is named in his honour.

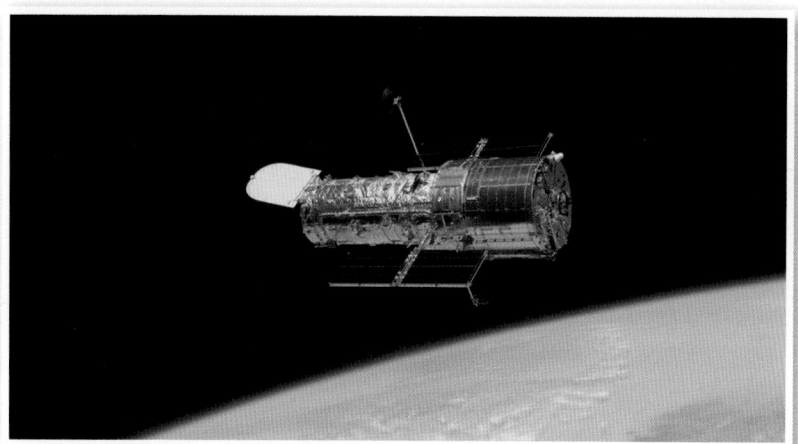

▲ Hubble Space Telescope

Questions

1 When a galaxy is moving away from us, in which direction is the light said to be shifted?

2 Describe what scientists have noticed about the red-shift from galaxies further away from us and explain what this means about those galaxies.

3 Why don't you notice a change in colour like you notice a change in pitch when race cars go past?

4 What would happen to the light observed from a galaxy that was moving towards us? Suggest a name for this effect.

Exam tip AQA

✔ It's important to remember not only that observations on red-shift show us that galaxies are moving away from us, but also that the galaxies furthest away are moving fastest.

32: The Big Bang theory

Learning objectives

After studying this topic, you should be able to:

✔ outline the Big Bang theory, including the evidence supporting it

✔ describe the origin of cosmic microwave background radiation

A Explain how the Universe was formed, according to the Big Bang theory, and describe what it was like in the past.

It all started with the Big Bang

The **Big Bang** is at present the most widely recognised scientific theory on how the Universe began. It states that the Universe began from a very small, very dense and very hot initial point. It burst outwards in a giant explosion, and all matter and space was created in the Big Bang. It is even thought that this was the moment when time began.

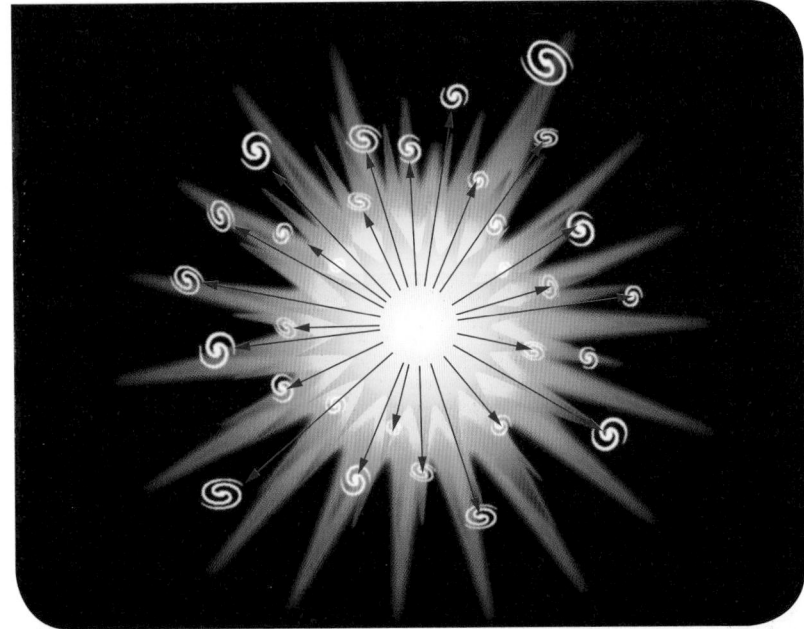

▲ Most scientists think the Universe was created in the Big Bang

This theory may seem a little strange, but there is some good evidence to support it.

Where's the evidence?

There are two key pieces of evidence for the Big Bang. The first is data collected on red-shift. You may remember that red-shift shows us that the galaxies which are furthest away from us are moving faster than galaxies closer to us. This suggests that the Universe is expanding outwards. The galaxies are a bit like the coloured sparks from an exploding firework. When the firework explodes, the sparks moving fastest travel the furthest.

If you could run time backwards, you would see the sparks coming back closer together and all starting in one point. The same is true for galaxies. They all started out at one point in space and then exploded outwards.

▲ An exploding firework

B What does red-shift suggest is happening to the Universe?

The second piece of evidence is even more compelling. In the 1960s, two scientists called Wilson and Penzias noticed that a form of microwave radiation was affecting their readings. No matter where they pointed their special telescope, they always detected the same background hum. This electromagnetic radiation was everywhere.

It is now called **cosmic microwave background radiation** (CMBR). Their explanations of this won Wilson and Penzias the Nobel prize for physics. They explained that CMBR must be the heat left over from the Big Bang. As the small, hot Universe expanded, it cooled and the radiation was stretched out. Today this radiation is in the microwave region of the electromagnetic spectrum. It is the same everywhere you look, because it fills the Universe. The Big Bang theory is currently the only theory than can explain the existence of this CMBR.

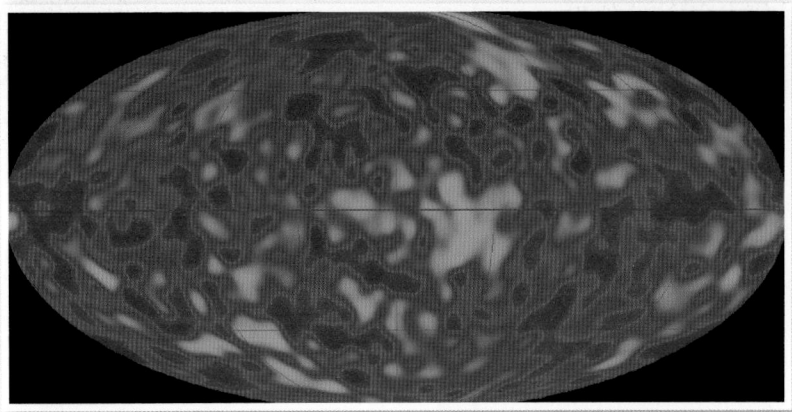

▲ Recent measurements show the distribution of cosmic microwave background radiation (CMBR)

Despite this **evidence** some scientists still disagree about the origin of the Universe. This is still one of the areas where future scientists will continue to explore other ideas. Who knows, maybe in a few years we will have a different theory. However, any change in the theory must explain the origin of CMBR.

Key words

Big Bang, cosmic microwave background radiation, evidence

Did you know...?

There are still a great number of unanswered questions in physics. For example, the scientific theories do not explain how or why the Universe began. Other recent observations suggest some very strange events at the edge of the Universe – it is not expanding in the way the Big Bang predicts. Scientists have come up with a number of temporary theories, including dark matter and dark energy, to try to explain these observations.

Questions

1 What are the two key pieces of evidence in support of the Big Bang theory? E

2 How is red-shift used to support the Big Bang theory?

3 Explain the origin of cosmic  C microwave background radiation.

4 Explain why the Big Bang theory is at present the most widely accepted theory on the origin of the Universe. A*

Course catch-up

Revision checklist

- ○ Some power stations use fossil fuels, biomass, and nuclear fuels to heat water. Steam then turns a turbine that drives an electrical generator.
- ○ Water in hydroelectric generators drives turbines directly. So does hot water and steam from geothermal areas. Solar cells convert the Sun's radiation energy directly into electricity.
- ○ Different energy sources affect the environment differently; releasing substances into the atmosphere, producing noise and visual pollution, producing waste, or destroying habitats.
- ○ Electricity is distributed from power stations to homes using the National Grid, using step-up and step-down transformers and high-voltage cables.
- ○ For a given power, increasing the voltage reduces the current required and this reduces the energy losses in the cables.
- ○ Waves transfer energy and can be either transverse or longitudinal. Waves have frequency, wavelength, and amplitude. They can be reflected, refracted, and diffracted.
- ○ Electromagnetic waves are transverse, sound waves are longitudinal. Mechanical waves can be either transverse or longitudinal.
- ○ Electromagnetic waves form a continuous spectrum. All types of electromagnetic waves travel at the same speed through a vacuum.
- ○ Radio waves, microwaves, infrared, and visible light can all be used for communication.
- ○ Longitudinal waves show areas of compression and rarefaction.
- ○ The normal is a line perpendicular to the reflecting surface at the point of incidence. The angle of incidence is equal to the angle of reflection.
- ○ The image produced in a plane mirror is virtual and upright.
- ○ Sound waves are longitudinal and cause vibrations in a medium, which are detected as sound. The pitch of a sound determines its frequency. Echoes are reflections of sounds.
- ○ The Doppler effect explains the change in wavelength and frequency of a wave seen by an observer. The Doppler effect on light from distant galaxies is called red-shift and is evidence for the Big Bang and expanding Universe.
- ○ The Big Bang is the only theory that currently explains cosmic microwave background radiation (CMBR).

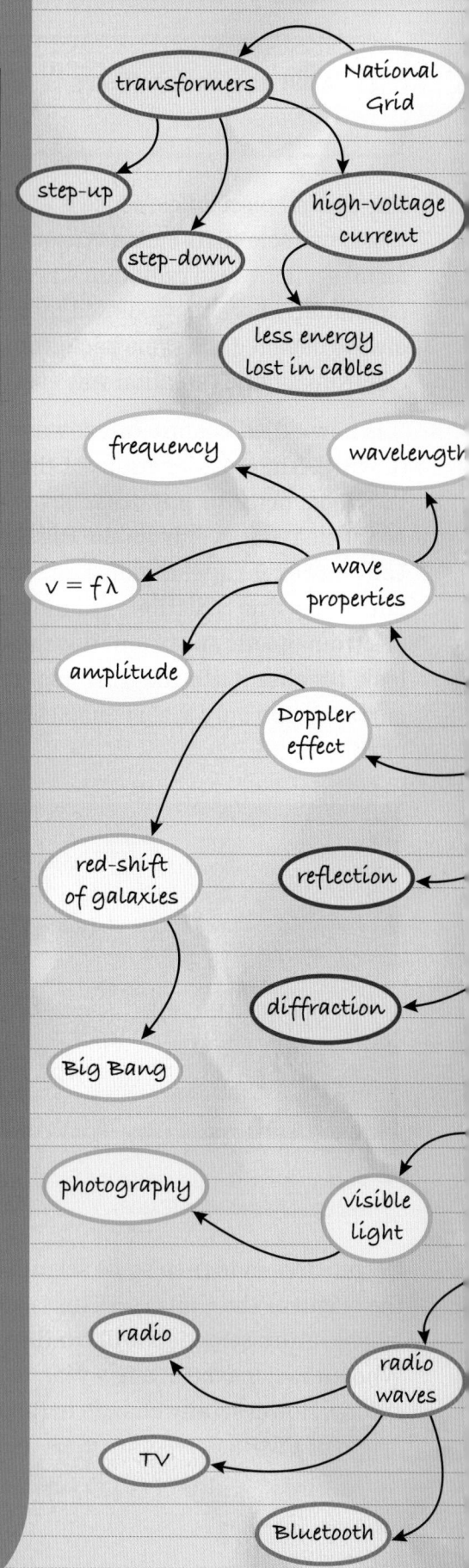

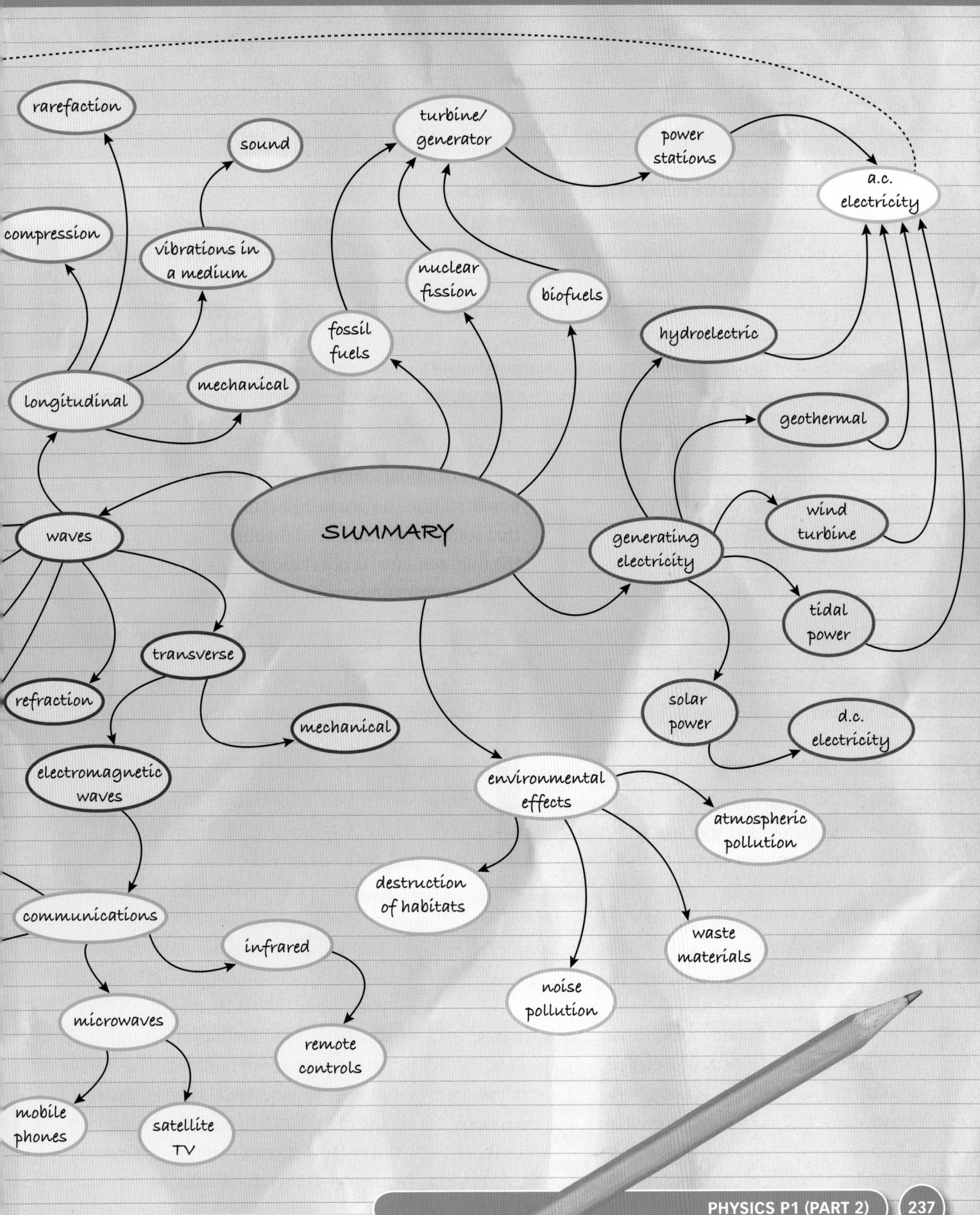

SUMMARY

rarefaction

sound

turbine/ generator

power stations

a.c. electricity

compression

vibrations in a medium

nuclear fission

biofuels

hydroelectric

longitudinal

mechanical

fossil fuels

geothermal

waves

wind turbine

transverse

generating electricity

tidal power

mechanical

refraction

solar power

d.c. electricity

electromagnetic waves

environmental effects

atmospheric pollution

communications

infrared

destruction of habitats

waste materials

microwaves

remote controls

noise pollution

mobile phones

satellite TV

Answering Extended Writing questions

QUESTION

William is an African schoolboy. He made a wind turbine from scrap materials to generate electricity for his home.

Outline the advantages and disadvantages of using wind turbines to generate electricity, compared to generating electricity in coal-fired power stations.

The quality of written communication will be assessed in your answer to this question.

G–E

Winde turbines do not make carbone dioxide, so there is no global warming from them My nan says that winde turbines in her village ruin the view from her bedroom. She is in a campaign against them.

Examiner: The points in the first sentence are correct, but the candidate then needs to mention that coal-fired power stations do produce carbon dioxide. It is true that some people find wind turbines unattractive, but the final sentence is not relevant. There are several spelling errors and one punctuation error.

D–C

Coal-fired power stations produce electrisity all the time, not just when it is windy, like wind turbines. Coal will run out, but wind is a reneweable resource. Birds can fly into wind turbines and then their lives come to a sad and sorry end, and if they pair for life, like doves, their partner will pine away.

Examiner: The candidate has made two good comparisons of the two methods of generating electricity, and used a scientific term correctly. The candidate could also have used the term 'finite resource' to describe coal. The final point is correct, but too detailed. The answer is well-organised, with two spelling errors.

B–A*

Coal is expensive, but wind is free, so it is cheaper to generate electricity from wind turbines (once you have built the turbines and the power station). Burning coal produces carbon dioxide, which causes global warming, and sulfur dioxide, which causes acid rain and asthma. Wind turbines do not cause pollution. But wind turbines kill birds and the noise is annoying. They only produce electricity when it is windy.

Examiner: This is a well organised answer that includes the key scientific points, and uses scientific vocabulary correctly. The spelling, punctuation, and grammar are faultless. The answer would have been even better if the candidate had added a comparison of coal-fired power stations with wind turbines in the final sentence.

Exam-style questions

1 State whether each of the following is a renewable or a non-renewable source of energy.

hydroelectricity

natural gas

solar

nuclear

biomass

coal

wind

tidal

oil

wave

A01

2 The illustration represents a gas-fired power station.

A01

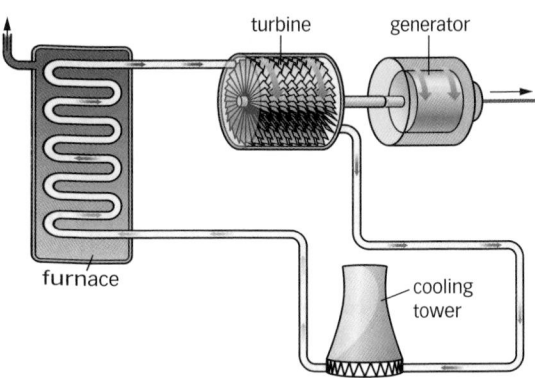

Explain what happens at each of the four stages illustrated above.

3 Light and radio waves travel at 3×10^8 m/s, and sound travels in air at 330 m/s.

A02

a Green light has wavelength 5×10^{-7} m. What is its frequency?

b The note middle C on the piano has frequency 256 Hz. What is its wavelength?

c What is the wavelength of an FM radio wave broadcasting at 100 MHz?

d A starting pistol is fired 200 m from where you are sitting. About how long is the interval between you seeing the smoke and hearing the bang?

Extended Writing

4 Describe some of the ways in which we use energy that reaches the Earth from the Sun.

A01

5 Explain what is meant by the Big Bang theory, and the main points of evidence that support it.

A01

6 There are many nuclear power stations operating successfully worldwide. Explain the process involved; and discuss some of the arguments for and against building more nuclear power stations.

A01

G–E

D–C

B–A*

G–E

D–C

B–A*

A01 Recall the science

A02 Apply your knowledge

A03 Evaluate and analyse the evidence

OXFORD
UNIVERSITY PRESS

Great Clarendon Street, Oxford OX2 6DP

Oxford University Press is a department of the University of Oxford.
It furthers the University's objective of excellence in research,
scholarship, and education by publishing worldwide in

Oxford   New York

Auckland   Cape Town   Dar es Salaam   Hong Kong   Karachi
Kuala Lumpur   Madrid   Melbourne   Mexico City   Nairobi
New Delhi   Shanghai   Taipei   Toronto

With offices in
Argentina   Austria   Brazil   Chile   Czech Republic   France   Greece
Guatemala   Hungary   Italy   Japan   Poland   Portugal   Singapore
South Korea   Switzerland   Thailand   Turkey   Ukraine   Vietnam

Oxford is a registered trade mark of Oxford University Press
in the UK and in certain other countries

© Oxford University Press 2011

The moral rights of the authors have been asserted

Database right Oxford University Press (maker)

First published 2011

All rights reserved. No part of this publication may be reproduced,
stored in a retrieval system, or transmitted, in any form or by any means,
without the prior permission in writing of Oxford University Press, or as
expressly permitted by law, or under terms agreed with the appropriate
reprographics rights organization. Enquiries concerning reproduction
outside the scope of the above should be sent to the Rights Department,
Oxford University Press, at the address above.

You must not circulate this book in any other binding or cover
and you must impose this same condition on any acquirer.

British Library Cataloguing in Publication Data

Data available

ISBN 978-0-19-913583-7

10 9 8 7 6 5 4 3 2 1

Printed in Great Britain by Bell and Bain, Glasgow

Paper used in the production of this book is a natural, recyclable product
made from wood grown in sustainable forests. The manufacturing process
conforms to the environmental regulations of the country of origin.

Mixed Sources
Product group from well-managed
forests and other controlled sources
www.fsc.org   Cert no. TT-COC-002769
© 1996 Forest Stewardship Council

FSC